"This book is inspirational. What William did took nothing more than initiative and a little learning, yet he changed his village and his life. There's never been a better time to Do It Yourself, and I love how much we can learn from those who often have no other choice."

—Chris Anderson, editor in chief of *Wired* and author of *Free* and *The Long Tail*

"I first met William onstage at TED. At the time, his English was faltering and he was understandably nervous. It didn't matter. His story, told in just a couple of minutes, was both astonishing and exhilarating. This book proves what those few minutes hinted at: a remarkable individual capable of inspiring many to take their future into their own hands."

—Chris Anderson, TED Curator

"A powerful read. This book takes you on a journey to discover pure innovation and the unfolding story of a natural genius. A true vision of struggle and tenacity to make a bold idea become a reality. This should be required reading for anyone who dares to dream."

—Cameron Sinclair, Eternal Optimist, Architecture for Humanity

"A moving, touching, important story. One more reminder of how small the world is and how powerful the human spirit can be."

—Seth Godin, author of *Tribes*

"Wonderful! I challenge you to read this story of one young man changing his corner of the world with nothing but intelligence and perseverance and not come away more hopeful about the prospects for a brighter, greener future."

—Alex Steffen, editor, Worldchanging.com

"Beyond opening the door to a nascent genre of African Innovation literature, *The Boy Who Harnessed the Wind* makes excuses about why Africans can't change their fates untenable. This potent, powerful, and uplifting message is the heart of William Kamkwamba's courageous story."

—Emeka Okafor, internationally acclaimed author of blogs
Timbuktu Chronicles and Africa Unchained

" In this book, the spirit, resilience, and resourcefulness that are Africa's greatest strengths shine through. My heart was gripped by the tale of how William's family pulled through the famine, and it was lifted up by the tale of how his determination brought light to his home and hope to his village. *The Boy Who Harnessed the Wind* is a remarkable story about a remarkable young man and his inquisitive and inventive mind."

—Amy Smith, founder, D-Lab, MIT

"I loved this enchanting story of a humble young hero from an impoverished African village who accomplished a miracle with scrap materials and unstoppable enthusiasm. What an inspiration!"

—Mark Frauenfelder, founder of boingboing.net; editor in chief of *MAKE*

"I was moved first to laughter, and then to tears by William's explanation of how he turned some PVC pipe, a broken bicycle, and some long wooden poles into a machine capable of generating sufficient current to power lights and a radio in his parents' house: 'I try, and I made it.'"

—Ethan Zuckerman, cofounder, Global Voices

"A rare and inspiring story of hope in rural Africa, a true story of youth challenging and winning against all of the adversity that life throws at it. William represents a new generation of Africans, using ingenuity and invention to overcome life's challenges. Where so many tilt at windmills, William builds them!"

—Erik Hersman, AfriGadget.com

"An inspiring tale of an African Cheetah—the new generation of young Africans who won't sit and wait for corrupt and incompetent governments—or vampire states—to come and do things for them. Here is one who harnessed the wind to generate electricity for his village—on his own."

—Professor George Ayittey, Distinguished Economist, American University

"William will challenge everything you have thought about Africa, about young people, and about the power of one person to transform a community. This beautifully written book will open your heart and mind. I was moved by William and his story and believe you all will. Essential, powerful, and compelling."

—Chris Abani, author of *Graceland*

"William Kamkwamba is an alchemist who turned misfortune into opportunity, opportunity beyond his own. The book is about learning by inventing. William's genius was to be ingenious."

—Nicholas Negroponte, founder, MIT Media Lab; founder and chairman, One Laptop per Child

"Much more than a memoir, this is a snapshot of life as a precocious teenager in contemporary Africa, and an affirmation of the notion that talent, beauty, and brilliance are distributed in equal measure around the world, even if opportunity is not. This is a story that hums with excitement of an individual who, like the continent where he was raised, is poised for greatness."

—Nathaniel Whittemore, Change.org

THE BOY
WHO HARNESSED
THE WIND

THE BOY
WHO HARNESSED
THE WIND

Creating Currents of Electricity and Hope

William Kamkwamba
AND BRYAN MEALER

WM
WILLIAM MORROW
An Imprint of HarperCollins*Publishers*

Map and illustrations in text by William Kamkwamba.

Photographs courtesy of: Kamkwamba family: pages 5, 26, 38; Bryan Mealer: pages 10, 57, 128–29; Tom Rielly, pages 185, 197, 209, 260; Sangwani Mwafulirwa, Malawi *Daily Times*: pages 238, 240

Designed by Jamie Lynn Kerner
Illustration of windmill on pp. ii–iii and chapter openers by Mary Schuck

Library of Congress Cataloging-in-Publication Data has been applied for.

ISBN 978-0-06-173032-0

09 10 11 12 13 OV/RRD 10 9 8 7 6 5

To my family

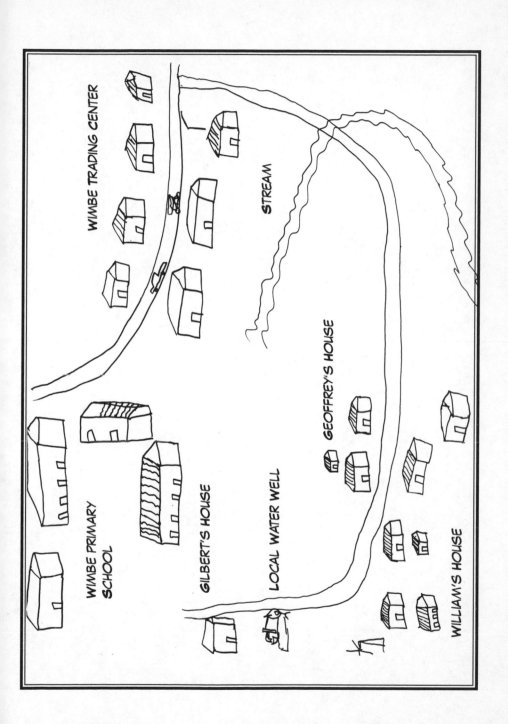

THE BOY

WHO HARNESSED

THE WIND

PROLOGUE

THE PREPARATION WAS COMPLETE, so I waited. The muscles in my arms still burned from having worked so hard, but now I was finished. The machinery was bolted and secured. The tower was steady and unmoving under the weight of twisted steel and plastic. Looking at it now, it appeared exactly as it was—something out of a dream.

News of the machine had spread to the villages, and people were starting to arrive. The traders spotted it from their stalls and packed up their things. The truckers left their vehicles along the roads. Everyone walked into the valley, and now gathered in its shadow. I recognized these faces. Some of these people had mocked me for months, and still they whispered, even laughed. More of them were coming. It was time.

Balancing the small reed and wires in my left hand, I used the other to pull myself onto the tower's first rung. The soft wood groaned under my weight, and the compound fell silent. I continued to climb, slowly and assuredly, until I was facing the machine's crude frame. Its plastic arms were burned and blackened, its metal bones bolted and welded into place. I paused and studied the flecks of rust and paint, how they appeared against the fields and mountains beyond. Each piece told its own tale of discovery, of being lost and found in a time of hardship and fear. Finally together now, we were all being reborn.

Two wires dangled from the heart of the machine and gently danced in the breeze. I knotted their frayed ends together with the wires that sprouted off the reed, just as I'd always pictured. Down below, the crowd cackled like a gang of birds.

"Quiet down," someone said. "Let's see how crazy this boy really is."

A sudden gust muffled the voices below, then picked up into a steady wind. It took hold of my T-shirt and whistled through the tower rungs. Reaching over, I removed a bent piece of wire that locked the machine's spinning wheel in place. Once released, the wheel and arms began to turn. They spun slowly at first, then faster and faster, until the force of their motion rocked the tower. My knees buckled, but I held on.

Don't let me down.

I gripped the reed and wires and waited for the miracle. Finally it came, at first a tiny light that flickered from my palm, then a surging magnificent glow. The crowd gasped and shuddered. The children pushed for a better look.

"It's true!" someone said.

"Yes," said another. "The boy has done it."

CHAPTER ONE

BEFORE I DISCOVERED THE miracles of science, magic ruled the world.

Magic and its many mysteries were a presence that hovered about constantly, giving me my earliest memory as a boy—the time my father saved me from certain death and became the hero he is today.

I was six years old, playing in the road, when a group of herd boys approached, singing and dancing. This was in Masitala village near the city of Kasungu, where my family lived on a farm. The herd boys worked for a nearby farmer who kept many cows. They explained how they'd been tending their herd that morning and discovered a giant sack in the road. When they opened it up, they found it filled with bubble gum. Can you imagine such a treasure? I can't tell you how much I loved bubble gum.

"Should we give some to this boy?" one asked.

I didn't move or breathe. There were dead leaves in my hair.

"*Eh,* why not?" said another. "Just look at him."

One of the boys reached into the bag and pulled out a handful of gumballs, one for every color, and dropped them into my hands. I stuffed them all in my mouth. As the boys left, I felt the sweet juice roll down my chin and soak my shirt.

The following day, I was playing under the mango tree when a trader on a bicycle stopped to chat with my father. He said that while on his way to the market the previous morning, he'd dropped one of his bags. By the time he'd realized what had happened and circled back, someone had taken it. The bag was filled with bubble gum, he said. Some fellow traders had told him about the herd boys passing out gum in the villages, and this made him very angry. For two days he'd been riding his bicycle throughout the district looking for the boys. He then issued a chilling threat.

"I've gone to see the *sing'anga,* and whoever ate that gum will soon be sorry."

The *sing'anga* was the witch doctor.

I'd swallowed the gum long before. Now the sweet, lingering memory of it soured into poison on my tongue. I began to sweat; my heart was beating fast. Without anyone seeing, I ran into the blue gum grove behind my house, leaned against a tree, and tried to make myself clean. I spit and hocked, shoved my finger into my throat, anything to rid my body of the curse. I came up dry. A bit of saliva colored the leaves at my feet, so I covered them with dirt.

But then, as if a dark cloud had passed over the sun, I felt the great eye of the wizard watching me through the trees. I'd eaten his juju and now his darkness owned me. That night, the witches would come for me in my bed. They'd take me aboard their planes and force me to fight, leaving me for dead along the magic battlefields. And as my soul drifted alone and forsaken above the clouds, my body would be cold by morning. A fear of death swept over me like a fever.

I began crying so hard I couldn't move my legs. The tears ran hot down my face, and as they did, the smell of poison filled my nose. It was everywhere inside me. I fled the forest as fast as possible, trying to get away from the giant magic eye. I ran all the way home to where my father sat against the house, plucking a pile of maize. I wanted to throw my body under his, so he could protect me from the devil.

"It was me," I said, the tears drowning my words. "I ate the stolen gum. I don't want to die, Papa. Don't let them take me!"

My father looked at me for a second, then shook his head.

"It was you, *eh*?" he said, then kind of smiled.

Didn't he realize I was done for?

"Well," he said, and rose from the chair. His knees popped whenever he stood. My father was a big man. "Don't worry. I'll find this trader and explain. I'm sure we can work out something."

Me as a young boy standing with my father in Masitala village. To me, he was the biggest and strongest man in the world.

That afternoon, my father walked eight kilometers to a place called Masaka where the trader lived. He told the man what had happened, about the herd boys coming by and giving me the stolen gum. Then without question, my father paid the man for his entire bag, which amounted to a full week's pay.

That evening after supper, my life having been saved, I asked my father about the curse, and if he'd truly believed I was finished. He straightened his face and became very serious.

"Oh yes, we were just in time," he said, then started laughing in that way that made me so happy, his big chest heaving and causing the wooden chair to squeal. *"William, who knows what was in store for you?"*

My father was strong and feared no magic, but he knew all the stories. On nights when there was no moon, we'd light a lamp and gather in our living room. My sisters and I would sit at my father's feet, and he'd explain the ways of the world, how magic had been with us from the beginning. In a land of poor farmers, there were too many troubles for God and man alone. To compensate for this imbalance, he said, magic existed as a third and powerful force. Magic wasn't something you could see, like a tree, or a woman carrying water. Instead, it was a force invisible and strong like the wind, or a spider's web spun across the trail. Magic existed in story, and one of our favorites was of Chief Mwase and the Battle of Kasungu.

In the early nineteenth century, and even today, the Chewa people were the rulers of the central plains. We'd fled there many generations before from the highlands of southern Congo during a time of great war and sickness, and settled where the soil was reddish black and fertile as the days were long.

During this time, just northwest of our village, a ferocious black rhino began wreaking terror across the land. He was bigger than a three-ton lorry, with horns the length of my father's arms and points as sharp as daggers. Back then, the villagers and animals shared the same watering hole, and the rhino would submerge himself in the shallows and wait. Those visiting the spring were mostly women and young girls like my mother and sisters. As they dipped their pails into the water, the rhino would attack, stabbing and stomping them with its mighty hooves, until there was nothing left but bloody rags. Over a period of months, the feared black rhino had killed over a hundred people.

One afternoon, a young girl from the royal Chewa family was stomped to death at the spring. When the chief heard about this, he became very angry and decided to act. He gathered his elders and warriors to make a plan.

"This thing is a real menace," the chief said. "How can we get rid of it?"

There were many ideas, but none seemed to impress the chief. Finally one of his assistants stood up.

"I know this man in Lilongwe," he said. "He's not a chief, but he owns one of the *azungu*'s guns, and he's very good at magic. I'm certain his magical calculations are strong enough to defeat this black rhino."

This man was Mwase Chiphaudzu, whose magic was so superior he was renowned across the kingdom. Mwase was a magic hunter. His very name meant "killer grass" because he was able to disguise himself as a cluster of reeds in the fields, allowing him to ambush his prey. The chief's people traveled a hundred kilometers to Lilongwe and summoned Mwase, who agreed to assist his brothers in Kasungu.

One morning, Mwase arrived at the watering hole well before the sun. He stood in the tall grass near the shores and sprinkled magic water over his body and rifle. Both of them vanished, becoming only music in the breeze. Minutes later, the black rhino thundered over the hill and made his way toward the spring. As he plunged his heavy body into the shallows, Mwase crept behind him and put a bullet into his skull. The rhino crumpled dead.

The celebrations began immediately. For three days, villagers from across the district feasted on the meat of the terrible beast that had taken so many lives. During the height of the festivities, the chief took Mwase to the top of the highest hill and looked down where the Chewa ruled. This hill was Mwala wa Nyenje, meaning "The Rock of the Edible Flies," named after the cliffs at its summit and the fat delicious flies that lived in its trees.

Standing atop the Rock of the Edible Flies, the chief pointed down to a giant swath of green earth and turned to Mwase.

"Because you killed that horrible and most feared beast, I have a prize for you," he said. "I hereby grant you power over this side of the mountain and all that's visible from its peak. Go get your people and make this your home. This is now your rule."

So Mwase returned to Lilongwe and got his family, and before long, he'd established a thriving empire. His farmland produced abundant maize and vegetables that fed the entire region. His people were strong, and his warriors were powerful and feared.

But around this time, a great chaos erupted in the Zulu kingdom of South Africa. The army of the Zulu king, Shaka, began a bloody campaign to conquer the land surrounding his kingdom, and this path of terror and destruction caused millions to flee. One such group was the Ngoni.

The Ngoni people marched north for many months and finally stopped in Chewa territory, where the soil was moist and fertile. But because they were constantly on the move, hunger visited them often. When this happened, they would travel farther north and ask for help from Chief Mwase, who always assisted them with maize and goats. One day, after accepting another of Mwase's handouts, the Ngoni chiefs sat down and said, "How can we always have this kind of food?"

Someone replied, "Eliminate the Chewa."

The Ngoni were led by Chief Nawambe, whose plan was to capture the Rock of the Edible Flies and all the land visible from its peak. However, the Ngoni did not know how magical Chief Mwase was.

One morning, the Ngoni came up the mountain dressed in animal skins, holding massive shields in one hand and spears in the other. But of course, Chief Mwase's warriors had spotted them from miles away. By the time the Ngoni reached the hill, the Chewa warriors had disguised themselves as green grass and slayed the intruders with knives and spears. The last man to die was Chief Nawambe. For this reason, the mountain was changed from the Rock of the Edible Flies to Nguru ya Nawambe, which means simply "The Deadly Defeat of Nawambe." This same hill now casts a long shadow over the city of Kasungu, just near my village.

THESE STORIES HAD BEEN passed down from generation to generation, with my father having learned them from my grandpa. My father's father was so old he couldn't remember when he was born. His skin was so dry and wrinkled, his feet looked like they were chiseled from stone. His overcoat and trousers seemed older than he was, the way they were patched and hung on his body like the bark of an ancient tree. He rolled fat cigars from maize husks and field tobacco, and his eyes were red from *kachaso,* a maize liquor so strong it left weaker men blind.

Grandpa visited us once or twice a month. Whenever he emerged from the edge of the trees in his long coat and hat, a trail of smoke rising from his lips, it was as if the forest itself had taken legs and walked.

The stories Grandpa told were from a different time and place. When he was young—before the government maize and tobacco estates arrived and cleared most of our trees—the forests were so dense a traveler could lose his sense of time and direction in them. Here the invisible world hovered closer to the ground, mixing with the darkness in the groves. The forest was home to many wild beasts, such as antelope, elephant, and wildebeest, as well as hyenas, lions, and leopards, adding even more to the danger.

When Grandpa was a boy, his grandmother was attacked by a lion. She was working in her fields at the forest's edge, scaring away some monkeys, when a female lion came upon her. Villagers heard her cries and quickly sounded the drum—not the fast, rhythmic beat for dances or ceremonies, but something slow and serious. They call this emergency beat the *musad-abwe,* meaning, "Don't ask questions, just come!" It's like dialing 911, but instead of police, you're calling other villagers.

By the time Grandpa and others arrived with their spears and bows and arrows, it was too late. They saw the lion—its body the size of a cow—drag his grandmother into the thorny trees, then toss her body into the bush like a mouse. It then turned and faced its challengers, let out a terrible roar, and disappeared with its kill. The poor woman's body was never recovered.

Grandpa says that once a lion gets a taste for human blood, it won't stop until it's eaten an entire village. So the next morning someone notified the British authorities, who still controlled our country. They sent soldiers into the forest and shot the lion. Its body was then displayed in the village square for all to see.

Not long after, Grandpa was hunting alone in the forest and came upon a man who'd been bitten by a cobra. The snake had been hiding in the trees and struck the man's head as he passed. His skin quickly turned gray, and minutes later, he was dead. Grandpa alerted the nearest village, who arrived with their witch doctor. The wizard placed one foot atop the dead man's chest and tossed some medicines into the forest. Seconds later, the moist ground came alive as hundreds of cobra slithered out from the shad-

Grandpa displaying his handmade bow and arrow, once used to kill lions and wildebeest. People say Grandpa was the greatest hunter in the district.

ows and gathered around the corpse, hypnotized by the spell.

The wizard crouched on the dead man's chest and drank a cup of magic porridge, which flowed through his feet and into the lifeless body. The dead man's fingers began to move, then his hands.

"Let me up," he said, then stood and faced the army of serpents.

Together, they checked the fangs of every cobra in attendance, searching for the one that had killed the man. Usually, the wizard would quickly cut off the head of the guilty snake, but this time, the dead man took pity and allowed the cobra to live. For his services, the wizard was paid three British pounds. My grandpa saw this with his own eyes.

When my father was a young man, he often went hunting with his father. Even then, the forest was so dangerous that hunters observed a sacred ritual before their outings. Hunts were usually initiated by one man, the *mwini chisokole,* or owner of the hunt, who called together all the willing men from the surrounding villages. The owner decided where and when the hunt would take place, and in the event of a kill, he'd receive the choicest portion of the meat, usually the hindquarter. Grandpa was often this person.

On the night before the hunt, the leader wasn't allowed to sleep with his wife, not even in the same room. The purpose was to keep the man's focus and attention as sharp as possible, and to guarantee a solid night's rest. Losing focus made you careless in the forest, and worst of all, left you open to bewitching. That night, sleeping alone at a neighbor's house, or in a separate hut with his sons, the leader would boil a pot of red maize mixed with certain roots and medicines, which he'd distribute the following morning to each hunter in the party. This was part of the magic, because everyone believed this protected them from danger.

Before setting out, the hunters also instructed their wives to stay indoors until the hunt was over, preferably lying in bed and sleeping. They thought this would cause the animals to sleep as well, allowing the hunters to sneak up on them with ease.

WALKING THROUGH THE FOREST as a boy, I didn't worry so much about cobras or lions, since most of them had vanished. But other dangers were waiting in the forests that remained, and along the quiet, empty fields where the ghosts of trees seemed to whisper their sadness. Walking there alone, one of my greatest fears was the Gule Wamkulu.

The Gule Wamkulu were a secret gang of dancers. They performed at the chief's request at funerals and initiation ceremonies, when many Chewa boys become men. The Gule Wamkulu were said to be the spirits of our dead ancestors, resurrected from the afterworld and sent to roam the earth. No longer human, they shared the skin of animals, and their faces resembled the beasts of hell—twisted devil birds and demons howling in fright.

When the Gule Wamkulu performed, you dared to watch only from a distance. Often they appeared from the bush walking on stilts, towering above the crowd and screaming in different tongues. Once, I even saw one of them climb a blue gum pole while upside down, like a spider. And when they danced, one thousand men seemed to inhabit their bodies, each moving in the opposite direction.

When the Gule Wamkulu weren't performing, they traveled the forests and marshes looking for young boys to take back to the graveyards. What happened to you there, I never wanted to know. It was bad luck to even speak about the Gule Wamkulu. And God help you if you were ever caught doubting them, saying, "Look at their hands, they have five fingers like me. These guys are not real." Doing this would surely get you bewitched, and since the Gule Wamkulu answered only to the chief, there'd be no one to defend you. When they appeared in the village, every woman and child dropped what they were doing and ran.

Once when I was very young, a magic dancer appeared in our courtyard, strutting like a cock and hissing like a snake. His head was wrapped in a flour sack with a black hole for a mouth and a long trunk for a nose. My mother and father were in the fields, so my sisters and I ran for the trees, only to watch this passing ghost steal one of our chickens.

(Donkeys are the only creatures not afraid of Gule Wamkulu. If the donkey sees one of these dancers, it will chase them into the bush and kick them with its mighty legs. Don't ask me why, but the donkey is very brave.)

I tried to be courageous like my friend the donkey whenever I walked through the forest. But witches and wizards never reveal their identity, so you never know where their traps lie waiting. In these places where they practice, their potent magic takes on many shapes. Men with bald heads, twenty feet tall, are said to appear on the roads outside of Ntchisi, a few at first, then dozens all around. Ghost trucks drive the same roads at night, coming on fast with their bright lights flashing and engines revving loud. But as the lights pass by, no truck is attached. No tire marks are left on the road, and if you're driving a car, your engine will die until morning.

Magic hyenas wander the villages at night, snatching several goats at once in their razor jaws and delivering them to the doorsteps of wizards. Magic lions are sent to kill delinquent debtors, and snakes the size of tractors can lie in wait for you in your fields.

But the dangers for children are even greater. As I mentioned, these wizards command great armies of children to do their witchcraft, and each night they prowl the villages for fresh recruits. They tempt them with delicious meats, saying it's the only way to heaven. Once the children devour the tasty morsels, it's revealed as human flesh. By then it's too late, for once the wizard's evil is inside your body, it controls you forever.

In addition to casting spells for curses and revenge, the witches often battle one another. This leads to great confusion in the kingdom of the devil, and this strife leaves many dead and injured, which is why children make the perfect soldiers.

The children pile aboard witch planes that prowl the skies at night, capable of traveling to Zambia and London in a single minute. Witch planes can be anything: a wooden basin, a clay pot, a simple hat. Flying about on magic duty, the children are sent to homes of rival wizards to test their powers. If the child is killed in the process, the wizard can determine the weapon of his enemy and develop something stronger. Other nights, the children visit camps of other witches for competition. Here, mystical soccer matches are played on mysterious fields in places I've never heard of, where the cursed children use human heads as balls and compete for great cups of flesh.

AFTER ESCAPING THE BUBBLEGUM vendor, I became terribly afraid of being captured, and I tried to think of ways to protect myself. I knew witches and wizards were allergic to money because the presence of cash is like a rival evil. Any contact with money will snap their spell and revert them back to human form—usually naked. For this reason, people often plaster their walls and bed mats with kwacha notes to protect themselves during the night. If they're suddenly awoken by a naked man trying to escape, their suspicions are correct.

Another way of protecting yourself is to pray your soul clean each night at the foot of your bed, and I'd done that, too. Homes of the prayerful are concealed from witch planes that fly overhead. It's like passing through a cloud.

"Papa, please, some kwacha notes for my walls," I begged my father one afternoon. "I can't sleep at night."

My father knew a lot about witchcraft, but he had no place for magic in his own life. To me, this made him seem even stronger. My parents had raised us to be churchgoing Presbyterians who believed God was the best protection. Once you opened your heart to magic, we were taught, you never knew what else you might let inside. We respected the power of juju, even feared it, but my family always trusted our faith would prevail.

My father was mending a fence around the garden and stopped what he was doing. "Let me tell you a story," he said. "In 1979 when I was trading, I was riding in the back of a pickup going to Lilongwe to sell dried fish in the market. Several others were with me. The truck suddenly lost control, pitching us all into the air. When we landed, we saw it rolling straight for us. I said at that moment, 'I'm dying now. This is my time.' But just before the truck rolled over my body and crushed me like an ant, it skidded to a stop. I could reach out and touch it. Several people were dead in the grass, but I didn't have a scratch."

He turned to face me, making his point.

"After that happened, how can I believe in wizards and charms? A magic man would have tried these things and died. I was saved by the power of God. Respect the wizards, my son, but always remember, with God on your side, they have no power."

I trusted my father, but wondered how his explanation accounted for Rambo and Chuck Norris, who came to the trading center that summer and created a lot of controversy. These men were appearing in films shown in the local theater, which was really just a thatch hut with wooden benches, a small television, and a VCR. For this reason, everyone called it the video show. At night, wonderful and mysterious things began happening in this place, but since I was forbidden to be out after dark, I missed them all. Instead, I relied on the stories I heard from my mates who lived close by and whose parents weren't so strict. These boys, such as Peter Kamanga, would find me the next day when I arrived.

"Last night I watched the best of all movies," Peter said. "Rambo jumped from the top of the mountain and was still firing his gun when he landed. Everyone in front of him died and the entire mountain exploded." He clutched a phantom machine gun and sent a burst of deadly rounds into the maize mill.

"Oh," I said, "when will they ever show these films during the day? I never see anything."

The exploits of Rambo and Delta Force became confusing to some, who'd never imagined men escaping entire armies, while still managing to kill so many people. The night *Terminator* came to the video show was simply shocking. When Peter found me the next morning, he was still in a state.

"William, last night I watched a movie that I still don't understand," he said. "This man was shot left, right, and center, yet he still managed to live. His enemies blew off his arms and legs, even his head, yet his eyes were still alive. I'm telling you, this man must be the greatest wizard who ever lived."

It sounded fantastic. "Do you think these *azungu* from America have such magic?" I asked. "I don't believe it."

"This is what I saw. I'm telling you it's true."

Although it would be several years before I finally saw one of these films in the video show, they started to influence many of the games we played back home. One of them was played with toy guns we made from a *mpoloni* bush.

It was called USA versus Vietnam.

To make these guns, we removed the core from the *mpoloni*'s stem, much like disassembling a ballpoint pen, and used it as both a ramrod and trigger. After removing the core, we chewed up bits of maize pith and shoved them down the barrel, followed by paper spitballs to create a seal. When the ramrod was forced down behind, it created enough pressure to spray an opponent with a shower of slimy gunk.

I was captain of one team, while my cousin Geoffrey was captain of the other. Along with some other cousins and neighbors, we split into teams of five, then hunted one another in the maize rows and across the dirt courtyard that separated our house from Geoffrey's.

"You go left, I'll go right!" I instructed my comrades one such afternoon, then scrambled on knees and elbows through the red dirt. We were never clean.

I spotted a bit of Geoffrey's trouser from around the corner of the house, so I snuck around the opposite way without spooking the chickens. Once I was clear, I bolted around the corner. It was an easy ambush.

"*Tonga!*"

I jammed the ramrod down the barrel and released a shower of white saliva and mush, spraying my cousin square in the face.

He fell to the ground, holding his heart.

"*Eh, mayo ine!* I'm dead."

Usually, whichever team won first got to be America the following round, since America always defeated Vietnam in the video show.

WE WERE A SOLID gang of three: myself, Geoffrey, and our friend Gilbert, whose father was the chief of our whole Wimbe district. Everyone called Gilbert's father Chief Wimbe, even though his real name was Albert Mofat.

When we got bored with playing USA versus Vietnam, Geoffrey and I went to find Gilbert. Going over to Gilbert's house always guaranteed a show, as the chief's work was never done. As usual, we found a line of truck drivers, market women, farmers, and traders waiting outside under the blue gum trees to share their concerns and grievances. Each held a chicken under one arm, or a small bit of cash in hand as a gift for their great leader. During these personal encounters with the chief, people addressed him as "*Charo,*" the ruler of all the land.

"*Odi, odi,*" said a farmer at the doorstep, meaning *hello, can I come in?*

The chief's messenger and bodyguard, Mister Ngwata, stood at the door in his short pants and army boots, dressed as a policeman. It was Mister Ngwata's job to protect the chief and filter all of his visitors. He also handled all the chickens.

"Come, come," he said.

The chief sat on the sofa, dressed in a crisp shirt and nice trousers. Chiefs usually dressed like businesspeople, never in feathers and hides. That's in the movies. Chief Wimbe also loved his cat, which was black and white but had no name. In Malawi, only dogs are given names, I don't know why. The cat was always in the chief's lap, purring softly as the *charo* stroked its neck.

"*Charo, Charo,*" the farmer said, bending to one knee and gently clapping his hands as a sign of respect. "We have an issue that requires your intervention. The land you granted me fifteen years ago is being encroached upon by my brother's son. I need you to help so there's no bloodshed."

"Very well," the chief replied. "Let me think about this and carry out some research. Come back on Sunday and I'll have an answer."

"Oh, *zikomo kwambiri, Charo.* Thank you, with respect."

We waited until the farmer left and approached Mister Ngwata.

"We're here to see Gilbert," we said as we passed through the door.

"Hmmph."

Gilbert was in his room with a tape deck singing to Billy Kaunda, who'd just been voted Malawi's best musician of the year. For a boy, Gilbert had a beautiful singing voice and would later record two albums in Blantyre. My voice sounded like one of the guinea fowl that screeched in our trees as it pooped, but I never let that stop me.

"Gilbert, bo?"

"Bo!"

"Sharp?"

"Sharp!"

This was our slang, strictly observed at every meeting. The word *bo* was short for *bonjour,* started by some chaps learning French in secondary school and wanting to show off. I don't know where "sharp" came from, but it was like saying, "Are you cool?" If you were feeling really good, you could even go a bit further:

"Sure?"

"Sure!"

"Fit?"

"Fit!"

"*Ehhh.*"

"Let's go to trading," I said, meaning the trading center. "I hear the drunkards were spilling out of Ofesi last night."

This was the Ofesi Boozing Centre, a forbidden and therefore fascinating place. Ofesi sat on the outskirts of the trading center, one of the last shops before the road opened toward Chamama town. Loud, thumping music always played inside the dark doorway, even at noon. It was where men with screwed-up eyes would appear in the doorframe, smoking cigarettes, then toss out empty cardboard cartons of booze to join the mountain of others in the dirt. Whereas most people saw garbage in those cartons, we saw treasure and possibility.

Although Geoffrey, Gilbert, and I grew up in this small place in Africa, we did many of the same things children do all over the world, only with slightly different materials. And talking with friends I've met from America and Europe, I now know this is true. Children everywhere have similar ways of entertaining themselves. If you look at it this way, the world isn't so big.

For us, we loved trucks. It didn't matter what kind of trucks: four-ton lorries that rumbled past from the estates kicking up dust, or the half-ton pickups that traveled back and forth to Kasungu town, just an hour's drive away, with passengers squeezed in back like a pen full of chickens. We loved all trucks, and each week, we'd compete to see who could build the biggest and strongest one. While my friends from America could find miniature trucks already assembled in their shopping malls, we had to build ours from wire and empty cartons of booze. Even still, they were just as beautiful.

The cartons discarded by the drunkards at Ofesi once held Chibuku Shake Shake, a kind of beer made from fermented maize that is popular in Malawi. It's sour tasting and contains bits of maize that settle at the bottom, requiring you to shake it up before enjoying, hence the name. Believe it or not, it's actually nutritious. I'm not a drinker myself, but I've been told it takes several cartons of Shake Shake to get a person drunk, so of course, people in Ofesi drink as many as possible before tossing them into the road.

After washing out the booze, these cartons were ideal for making the chassis of a toy truck. We used beer bottle caps for wheels, which also doubled as counters at school ("Three Coca-Cola plus ten Carlsberg equals thirteen").

We picked mangoes from the neighbor's trees and traded them for lengths of wire, which we used to make axles and attach the bottle cap wheels. We later discovered that plastic cooking oil caps worked much better as wheels, enabling the trucks to last much longer. We even took our fathers' razor blades and cut designs into the plastic to give each vehicle its own signature treading. That way, when you saw a tire track in the dirt, you knew instantly if it belonged to the great fleets of Kamkwamba Toyota, for instance, or Gilbert Company LTD.

Soon we were building our own monster wagons, called *chigirigiri*, that resembled something like a go-cart in America. The frames were thick tree branches forked at one end, where a person could sit at the junction. We then dug up large, round tuber roots called *kaumbu* and carved them into wheels, using blue gum poles as axles. All the loose parts were then lashed together with vine and tree bark.

Taking a rope, one person pulled the car while the driver steered with his feet. With two cars side by side, we held monster derbies down the dirt road.

"Let's race."

"For sure."

"Last one to reach Iponga's will go blind!"

"GO!"

Iponga Barber Shop was the first of its kind in the Wimbe trading center, and where I got all my haircuts. When my father brought me there each month, Mister Iponga would drape me in a tattered sheet and say, "What will it be?" Pictures of men with many different styles hung from the wall—styles such as the Tyson, after the famous American boxer, in addition to the English Cut, the Nigeria, and the Buddha, which was totally bald. I usually went for the Office Cut, which was close all over without any frills. I think it was the cheapest, too.

Of course, the problem with getting haircuts in the trading center was the frequent power outages that plagued the country. These could easily happen while Iponga held his electric clippers to your head.

"Oops, lost the power. Come back in a few hours."

"But . . ."

The best idea was to bring a hat, or else go at night, so you could slip home under the cover of darkness and return the next morning to have it fixed.

If we had some pocket change from our parents, we stopped by Mister Banda's shop for a cold bottle of Fanta or a handful of Dandy sweets, which Banda kept in a glass jar below the shelves of Drews liver salts and Con Jex cough tabs, Top Society Luxury lotions, Easy Black hair dye, long ribbons of Blue Band margarine, bars of Lifebuoy soap, and packets of Cowbell powdered milk.

Or if we were hungry, we pooled our money and headed to the *kanyenya* stand, which was really just a giant vat of boiling grease over a fire, next to the boozing center. There we bought delicious pieces of fried goat and chips for just a few kwacha. The man working the vat grunted, "How much?" and you answered, "Five kwacha." He sawed off a good chunk from a carcass hanging on the gallows, causing the swarm of black flies to circle once, then land again. He dropped the meat into the oil, added a few more sticks to his fire to get a raging boil, then threw in a handful of sliced potatoes. When everything was finished, he tossed them onto the counter, along with a small pile of salt for dipping.

"You mother is a good cooker," said Gilbert. "But she's never made anything as good as this."

"For sure."

But most of the time we had no money, so we spent our afternoons in hunger and dreams. On our way home we played a certain game with the *mphangala* bush. Its bright red flowers made the perfect crayons for children, but its stems could also tell your fortune. One person uprooted the stem, then tried to split it down the middle by pulling it apart. If you did this without breaking the stem in half, you'd have meat for dinner waiting for you at home.

"*Eh* man, you're lucky. Let me come over!"

But if you broke the stem, that was a different story.

"Oh, sorry, friend, your mother's at a funeral. You'll find only water at home! HA! HA!"

Evenings in the village, just after the sun disappeared over the blue gums, were my favorite time of day. This was when my father and Uncle

John—Geoffrey's father—finished work in the maize and tobacco fields and returned home for supper. My mother and older sister Annie would be busy in the kitchen preparing the food, sending out all the delicious smells riding on the breeze. All my cousins would gather in the courtyard between my house and Geoffrey's house to kick the soccer ball—made from plastic shopping bags we called *jumbo*s, which we then bound in twine. And as the light faded, perhaps a farmer from the next village would stop by.

"Mister Kamkwamba, I have something from my garden," he'd say, opening a bundle of papers to reveal some nice tomato plants. They'd negotiate a price and my father would plant them behind the house.

During the rainy season when the mangoes were ripe, we filled our pails with fruit from the neighbor's trees and soaked them in water while we ate our supper. Afterward, we passed the fruits around, biting into the juicy meat and letting the sweet syrup run down our fingers. If there wasn't any moonlight to continue playing, my father gathered all the children inside our living room, lit a kerosene lamp, and told us folktales.

"Sit down and hush up," he said. "Have I told the one about the Leopard and the Lion?"

"Tell it again, Papa!"

"Okay, well . . . one day long long ago, two girls were walking from Kasungu to Wimbe when they became too tired to continue."

We sat on the floor, hugging our knees against our chests and hanging on every word. My father knew many stories, and the Leopard and the Lion was one of my favorites. It went like this:

Rather than taking a nap in the dirt, the two young girls looked for a clean, quiet place to sleep. After some time, they came across the house of an old man. After making their request, the old man said, "Of course you can stay here. Come on in."

That night when the girls were fast asleep, the old man snuck out the door and walked into the dark forest. There he found his two best friends, the Leopard and the Lion.

"My friends, I have some tasty food for you. Just follow me."

"Why thanks, old man," the Leopard said. "We're coming straightaway."

The old man led his two friends through the forest and back to his house. The Leopard and the Lion were so excited for their meal they even started singing a happy tune. But as they were approaching, the two girls happened to wake up. They felt refreshed after their nap and decided to continue on their journey. Not seeing the old man, they left a kind note thanking him for the bed.

Finally, the old man arrived at the house with the Leopard and the Lion.

"Wait here and I'll go and get them," he said.

The old man saw the bed was empty. Where did they go? he wondered. He looked for the girls but couldn't find them. Finally, he discovered the note and knew they were gone. Outside, the Leopard and the Lion were growing impatient.

"Hey, where's our food?" said the Leopard. "Can't you see we're salivating out here?"

The old man called out, "Hold on, they're here someplace. Let me find them."

The old man knew if the Leopard and the Lion discovered that the girls had gone, they would surely eat him for supper instead. The old man kept a giant gourd in the corner of his house for drinking water. Seeing no other option, he jumped inside and hid.

Finally, after waiting so long, the Lion said, "That's it. We're going in!"

They broke open the door and found the house empty. No girls, no old man, no supper.

"Hey, the old man must've tricked us," said the Leopard. "He's even left himself."

Just then, the Leopard spotted a bit of the old man's shirt hanging out from the gourd. He motioned to the Lion, and together they tugged and tugged until the old man came flying out.

"Please no, I can explain," cried the old man. But the Leopard and the Lion had no patience for stories and quickly ate him.

My father clapped his hands together, signaling the end of the story. Then he looked around to all of us children.

"When planning misfortune for your friends," he said, "be careful because it will come back to haunt you. You must always wish others well."

"Tell another, Papa!" we shouted.

"Hmm, okay . . . what about the Snake and the Guinea Fowl?"

"For sure!"

Sometimes my father would forget the stories halfway and make them up as he went along. These tales would spiral on for an hour, with characters and motives ever changing. But through his own kind of magic, the stories would always end the same. My father was a born storyteller, largely because his own life had been like one fantastic tale.

CHAPTER TWO

W HEN MY FATHER, TRYWELL, was a young man, HE was quite famous. These days he's a farmer, just like his own father and the father before him. Being born Malawian automatically made you a farmer. I think it's written in the constitution somewhere, like a law passed down from Moses. If you didn't tend the soil, then you bought and sold in the market, and before my father gave himself to the fields, he led the crazy life of a traveling trader.

This was when he lived in Dowa, a small town southeast of Masitala perched high in the brown hills. Back during the '70s and '80s, Dowa was a vibrant place where a young man could go and make some money. At that time, Malawi was under the control of Hastings Kamuzu Banda, a powerful dictator who ruled the country for more than thirty years.

Every Malawian grew up knowing the story of Banda. When he was a young boy in Kasungu, living in the shadow of the great mountain where the Chewa defeated the Ngoni, Banda had walked barefoot one thousand miles to work in the gold mines of South Africa. Later, he was given a scholarship to universities in Indiana and Tennessee, where he earned a degree in medicine. He was a doctor in England before he returned to Malawi to deliver us from British rule. He became our first great leader, and in 1971, under his extreme pressure, our Parliament gave him the title Life President.

Banda was a tough man. He demanded that every trader in Malawi hang his picture in his shop, and no other photo could dare hang higher. If you didn't have the image of our Dear President on the wall—dressed in his three-piece suit and clutching a flywhisk—you would pay a hefty price. It was a frightening and confusing period in our history. Banda also forbade women to wear pants or dresses above the knee. For men, having long hair would get you tossed in jail. Kissing in public was also forbidden, as were films where kissing was portrayed. The president hated kissing, and even today, people are scared of smooching in the open. On top of that, policemen and the Young Pioneers—Banda's personal thugs—were always snatching up people who dared criticize his policies. Many Malawians were jailed, tortured, and even tossed into pits of hungry crocodiles.

Despite all of this, it was an exciting time to be a trader. My father tells stories about hitchhiking in pickups across the countryside to Lake Malawi, where he bought bundles of dried fish, rice, and used clothing, to sell back in the Dowa market. Lake Malawi is one of the biggest in the world and nearly covers the entire eastern half of our country. It's so vast it has waves like an ocean. I was twenty years old before I ever saw this lake with my own eyes, despite having grown up only two hours from its shores. But once I stood on its banks and looked out across its endless-looking water, my heart was filled with a great love for my country.

Once at the lake, the traders would travel to the cities of Nkhotakota and Mangochi aboard the steamer ships *Ilala* and *Chauncy Maples*, where good food was served, and traders drank and danced on the decks through the voyage. At the lake my father bartered with the Muslim businessmen, known as the Yao, who populate that part of the country.

The Yao arrived in Malawi more than a hundred years ago from across the lake in Mozambique. The Arabs from Zanzibar convinced them to become Muslim, then recruited them to capture our Chewa people and put us into bondage. They raided our villages, killed our men, then sent our women and children across the lake in boats. Once there, the slaves were shackled by the neck and made to march across Tanzania. This took three months. Once they reached the ocean, most of them were dead. Later on,

The Pope in his crazy days, sitting at his stall (center with dark shirt) in the Dowa market with his pals.

the Yao captured and traded us to the Portuguese in exchange for guns, gold, and salt.

If it weren't for the great Scottish missionary David Livingstone, the Yao and Chewa might still be at odds today. Livingstone helped end slavery, opened Malawi to trade, and built good schools and missions. Young men became educated and earned money, and once these economic opportunities were available to all, our two tribes had little reason to fight. Today we consider the Yao our brothers and sisters. My mother herself is a Yao, and I am half Yao.

My father has told me many stories about the small town of Mangochi, located on the southern tip of the lake, just near the mouth of the Shire River. The way he describes this place makes it sound like the great bazaars of northern Africa I've read about in books. The streets were filled with traders from all over Malawi, Zambia, Tanzania, and Mozambique, all their different languages and songs mixing with the smell of sweating bodies, spices, fried fish, and roasted maize. Pocketfuls of money were quickly emptied in the boozing dens, and by professional ladies of the night, who lured traders into their rooms for hot baths, expensive food, and other pleasures I didn't understand until I was older. Often, traders got carried away in such places and ran out of money. My father remembers seeing men running away with nothing but their underpants.

Many of these same traders also had wives and children back home, in addition to the prostitutes. This was well before my father met my mother, back when he was young and too busy traveling to be tied down with a woman or family. He had a few girlfriends, sure, but he generally stayed away from the bar girls. And because of his reluctance to do this, the people in the market started calling him the Pope.

"*Eh, Papa*," they'd tease, using the Chichewa word. "What happened? Did you fall off the pawpaw tree and break your testicles? Don't listen to your mother—these girls don't really burn!"

My father endured this teasing, because what else could he do? And after a while, that name caught on with so many people that hardly anyone remembered where it came from.

MY FATHER WAS A giant man, but his tolerance for alcohol was even greater. One night he and his friends settled down in the Dowa General Grocery at 5:00 P.M. As my father tells it, he drank fifty-six bottles of Carlsberg beer, and at 2:00 A.M. walked home to tell the story. These drinking sessions sometimes led to fistfights, which my father welcomed like sport.

After a while, he became one of the most famous traders around, but not just for his cleverness in business, or his ability to drink crates of beer. My father was legendary for his strength. In Malawi we like to say, "One head cannot lift up the roof." Well, my father must not have been listening.

Every July 6, we Malawians celebrate our independence from England, much like our brothers and sisters do in America on July 4. And like in the United States, the way we celebrate is with great parties filled with lots of music, dancing, and delicious grilled meats. It was on such a holiday that Robert Fumulani, the holy father of Malawian reggae music, came to sing at Dowa District Hall, and my father—then twenty-two years old—was determined to go.

Robert Fumulani was my father's most favorite singer. Fumulani's songs often described the struggles of the poor, his lyrics straight from the warm red Malawian soil. My father had seen Fumulani perform many times already, in Kasungu, Lilongwe, Nkhotakota, and Ntchisi, and each time, the singer wore his signature white shirt that made him look sharp.

Well, if you can imagine, the line to see Fumulani on Independence Day began forming early, right around the time my father stepped up to the bar at General Grocery. Hours passed, and by the time he stumbled outside, the beautiful sounds of Fumulani's voice could be heard all over town. The concert had begun.

My father rushed over to the hall, where he found a line still waiting to get inside. If you've ever stood with us Africans at airports or bus depots, you know we're never good with lines. What if we miss something? So wasting no time, my father pushed his way to the front, but was stopped at the door by a policeman.

"The concert is full," the policeman announced. "No one else allowed inside."

My father presented his ticket, but the policeman still refused. Being a bit drunk and bold, my father pushed the policeman aside and quickly mixed into the crowd. Once there, he discovered what a great party it was! There onstage was Robert Fumulani and his Likhubula River Dance Band, with the singer dressed in his smart white shirt and his guitar strapped to his neck. In the back, workers tended to giant barbecue and *kanyenya* stands loaded with delicious goat and beef. And of course, there was lots of Carlsberg.

Overcome with excitement, my father squeezed through the mob of sweaty bodies until he reached the front. Fumulani was singing one of his most beloved songs, "Sister," about his estranged wife.

"Lady," he sang, "don't insult me today just because I'm poor. You don't know what my future holds . . ."

As if hypnotized by this wonderful music, my father began to dance. But he wasn't doing just any dance—he was a man *possessed*, a man who knows in his heart that he is the greatest dancer on earth. His arms and legs became as graceful as a gazelle's, and his giant body sprang in the air like a flying grasshopper. Oh, what moves! But when he opened his eyes, he realized the music had stopped. Everyone on the floor now stood in silence. Robert Fumulani, the blessed father of our national music, stared down, looking angry.

He pointed to my father and called out, "Someone remove this drunkard from the floor. He's ruining my show!"

The crowd shouted and hissed, "He is here! Take him away!"

My father was crushed. How could this be? He was just having a good time, and now he was being called down like a child by our dear hero. Feeling betrayed, he straightened himself and pointed to the stage.

"Mister Fumulani," he yelled, "I have an invitation to be in this room. And like every Malawian here celebrating their proud independence, I am doing the same. I'm not the only person here who is drunk, you know. Besides, isn't it your job to sing and entertain?"

A line of policemen and Young Pioneers now circled the dance floor, waiting to pounce.

"Mister Fumulani, I only wish to dance in peace," my father said, then turned to face the police. "But since you've asked these men to remove me, *I say let them come!*"

The policemen swooped in and swallowed my father in a swarm of fists and elbows. The crowd rushed in behind. From the look of things, it appeared my father had been properly handled.

But suddenly, one by one, the policemen began flying off the pile as if wrestling a cyclone. They twisted in the air like sacks of flour and limped off in pain. When the last policeman was pitched to the wall, the room erupted in cheers.

There stood the Pope in the center of the crowd, shaking his mighty fists.

"*Who is next?*" he shouted. "*I'LL FIGHT YOU ALL!*"

A pack of Young Pioneers then tried their luck, only to be pitched off the same way. For half an hour, the cops and government thugs tried everything to shackle my father's hands, and each time, they failed. Too exhausted to continue fighting, my father finally agreed to be arrested and spend the night in jail ("Only because I respect the rule of law," he told them). However, he had one condition: that first he be allowed to enjoy his Independence Day barbecue. So after devouring a plate of delicious *kanyenya,* the Pope washed his hands and walked out with the police.

And that is the story of how my father fought twelve men and won.

Soon the story spread across the district and my father became famous. People congratulated him in the bars and markets of the lakeshore, and business improved as a result. This fame also attracted many of the thieves and robbers who lurked in the markets. "You're so strong," they said, slapping him on the back. "Let us use your strength to make us all rich!"

But my father was no criminal. He just wanted to work hard for his money and drink his Carlsberg. However, if anyone wished to fight, that could be arranged.

ALTHOUGH HIS FRIENDS HAD no idea, for quite some time the Pope had been keeping his eye on a particular girl. She appeared at the market at the same time each morning, only to disappear in the crowds. An hour would pass, and she'd reappear, carrying a bundle of vegetables or bag of flour, then make her way home to the neighborhood down the hill. These brief moments became the most important part of my father's day, and he made sure he was always at his stall where he could watch her. Even though he'd never heard her voice, something about her seemed to change something inside him. This girl, as you probably guessed, was my mother, Agnes.

Well, my father must not have been very smooth, because my mother was well aware of him staring, the way he gazed at her like a puppy at the henhouse door, never sure what to do. She'd asked around and knew his reputation. For some odd reason, these stories of fighting and misbehaving made her excited. Each day she couldn't wait for her mother to send her to the market. Even before entering the rows of wooden stalls, her heart would pound like the *chiwoda* drums of her childhood dances. Making her way across, it took everything inside her to keep from grinning. But my mother couldn't let on; she was no easy fish to catch.

This game of staring continued for several months, and my mother wondered if this man would ever make his move. If he was so strong and brave, then why on earth was he frightened of her? (As my father tells it, she was always too far away to chase after, and also, yes, he was terrified.)

Finally, my mother decided to test this big, powerful man.

One morning, my father saw her enter the market, and as usual, he quickly became lost in the sight of her. But this time she did something different. She took a new route through the market—one that was bringing her straight in his direction.

My father became nervous, but knew the time was now or never. This is my big chance, he thought, *but what will I say?* He didn't have time to think, because in a matter of seconds, my mother was right upon him. It was the closest she'd ever been, and the sight of her skin made his heart go mad, as if it was trying to run away.

Somehow, he found his courage and leaped over his stall. As she passed, he shouted, "You're the most beautiful woman I've ever seen!"

My mother spun around. My father was standing there in the row, arms open, those same eyes now meeting hers.

"I've loved you my entire life," he said. "And I want to marry you."

Struggling to stay composed, my mother said, "I'll have to think about that one," then turned and ran away.

Well, my father didn't give her much time. That very afternoon he was at her house, asking again. The next day, the same thing. My mother's older brother Bakili warned her about my father. Bakili was also a trader in the market and knew my father's reputation.

"He's always in the bars, drinking and fighting," he said. "Sister, this man is not a good husband."

"I don't care," my mother said. "He's so strong, and I love him."

Bakili then told their parents. My grandmother Rose was a tough woman, so tough she'd built the family home with her own hands while my grandpa worked as a tailor in the market. She'd even built the furnace and molded the bricks herself, which is not an easy job, and even today, not the job of a woman.

Hearing the news, my grandmother and grandfather confronted my mother.

"Now tell us the truth, Agnes. Are you serious about this man?"

"Yes," my mother said. "Double serious."

As it turned out, my grandfather had proposed to my grandmother in much the same way, after seeing her dance in a village competition. "The way she was dancing just stole my heart," my grandfather said. "And I said to myself, 'I'm going to marry her.'" He'd sent a young village girl to inform my grandmother he wanted to speak with her, only to have my grandmother confront him personally.

"You want to talk to me?" she said. "Then talk to me. What do you want?"

"For you to be my wife," he answered.

So what could my grandparents really say now? Six months later, Agnes married my father, and the following year, my sister Annie was born. But even with all these new developments, my father remained the Pope.

Well, the Pope's drunken lifestyle soon began to take its toll. My mother grew increasingly tired of him coming home drunk and smelling of booze, and often they'd argue. It was a dark period all around, a time that saw several of my father's closest friends die or go to prison, while others simply vanished.

First his friend Kafu picked up gonorrhea, known as the "bombs," from a prostitute in the bars. The veins that led to his testicles became swollen and rotten. One day, they exploded and Kafu died. Another friend named Mwanza was beaten to death in the pub over a girl. The new prostitute in town had made the mistake of flirting with both Mwanza and his friend. Well, they couldn't decide who was taking the lady home at the end of the night, so they decided to fight. It began innocently, but before anyone knew it, Mwanza was dead in a pool of blood. Of course, the prostitute fled before the first punch and never returned.

In Dowa, there was a famous preacher named Reverend JJ Chikankheni, who happened to be one of my father's most loyal customers. Reverend JJ led one of the biggest Presbyterian churches in Dowa, along with twenty-five smaller prayer houses across the district. He'd often stop by my father's stall and buy a bag of rice and the two men would chat. One day, the reverend looked deep into my father's eyes, as if scraping the bottom of his soul.

"Kamkwamba?" he said.

"Yes?"

"Do you know that God loves you, and that you disappoint Him every time you drink and fight and cause trouble?"

"Thanks, Reverend, but . . ."

"The good news is that even though you disappoint Him, He's ready to receive you. He wants you to turn to Him."

"Thanks, Reverend," my father said, trying to be polite. "Whatever you say."

A few nights later, my father was drinking as usual in the pub when a man walked up and knocked over his beer. The man was drunk and looking to fight the biggest guy in the room. Well, my father gave him what he wanted, and more. In a matter of seconds, the man lay on the floor with blood gushing from his ears. My father had to be pulled off the man, having nearly beaten him to death. The police soon arrived and arrested my father.

"You've really done it this time," the officer told him.

The head prosecutor in Dowa was a church deacon named Mister Kabisa, who was also one of my father's loyal customers. When Kabisa heard my father was in jail awaiting a trial, he paid a personal visit.

"Kamkwamba," he said, "I've always advised you not to indulge in these unnecessary fights. Someday you'll be killed or kill someone else, and look what happened here. You're my friend, and I don't want to lose you.

"You're supposed to go to court today and stand trial," Kabisa continued. "You'll probably lose and be sent to jail, perhaps even Zaleka prison. You've heard about the conditions there. Chances are you won't make it out alive."

Mister Kabisa then leaned in close and looked into my father's eyes the same way Reverend JJ had done, as if searching the dark corners of his heart.

"But I don't want you to go to prison. There's a better path for you. I'm willing to tear up these files and release you, but you have to promise me one thing."

"Anything," my father said.

"Turn your life over to God."

Of course, my father happily agreed just to get out of jail. But what the man said stayed in his mind. All that evening and the following day, it never gave him peace.

The following night while asleep, my father was visited by a dream. All he saw was darkness, nothing but an endless expanse of black. He felt confused and scared. It was as if he'd gone blind and couldn't shake himself awake. Then came a voice, piped in like a loudspeaker from heaven. It said: "These things will destroy you. Turn to me."

When my father awoke in the morning, his entire body was trembling like a baby bird's. The dream, plus all the advice and warnings of the past week, seemed too great a message to ignore. He woke up my mother, who lay sleeping beside him, and said, "My wife, today I'm turning to God. I've seen the signs, and now it's time to change."

That same morning, instead of going straight to work, my father stopped by the church to see Reverend JJ. The preacher was in his office.

"I'm here," my father said. "I'm ready."

My mother didn't recognize this new man who began coming home each night after work, this man who suddenly had lots of money for food and medicine for his kids. She was so happy, but still couldn't believe her good fortune. Each night for weeks, she'd still say, "Come here!" when he walked in the door, just to sniff his breath.

WHILE MY FATHER HAD been traveling, trading, and boozing, his older brother John had built up a booming business. Back in the late '60s and early '70s when President Banda was building all the big estates near Wimbe and Kasungu, there was lots of work for the local men. Building contracts were like gold, and Uncle John happened to know some of the managers who were hiring these subcontractors. Working as a kind of headhunter, John became the middleman, finding the right skilled, trustworthy crews to do the jobs. Because his judgment was always good, the estates paid him handsomly.

After several years of working for the estates, Uncle John saved enough money to start a farm imports business, buying and selling maize seed and fertilizer to the local farmers. He even had a small storefront in the trading center. This business became successful, and after a few years, he sold it and bought fifty-nine acres of land from Chief Wimbe, which he used to grow maize and burley tobacco—a kind of mild tobacco that's cured in the open air under handmade shelters.

Since Uncle John had money for good fertilizer, the tobacco from his farm was top quality. His fields never had any weeds and the leaves were deep green while growing, drying like the color of milk chocolate with fine traces of red. His tobacco fetched a high price each year at the Auction Holdings Limited in Lilongwe, where the farmers sold their hundred-kilogram bales on the auction floor. One good bale of tobacco would pay for seventeen more bags of fertilizer, enabling his farm to stay strong, given the good weather.

In 1989, when I was one year old, Uncle John came to Dowa for a friend's engagement party and stopped by for a visit. That night he and my father went for a walk.

"Why don't you come back to the village and farm with me," John said. "Things are going well."

"I can see," said my father. "But farming takes too long. I've gotten so used to the trading. How can I start something new?"

"It takes a long time, true. But if you invest that time and just a little money, the payoff is huge. Look what I'm making from tobacco. That kind of profit is impossible with trading. How much are you clearing each month with your rice and secondhand clothing? Five percent?"

"Four percent," my father said. "Soon I won't even be able to feed these kids. If I eat, my business suffers."

"Well, come back home, young brother. There's a big place waiting for you."

My father then told John he'd stopped drinking and turned his life to God.

"Then, think of this as a chance to start over," he said. "Consider this a sign."

"Okay," my father said. "You've convinced me."

By NOW WE HAD three kids (my sister Aisha had been born not long before) and my father saw this as an opportunity he couldn't resist. A few weeks later, after selling his stall in the market, he strapped all our belongings—our clothes, pots, pans, and the family radio—to the top of a UTM (United Transport Malawi) bus. We traveled four hours north to the Wimbe trading center, where my relatives were waiting to greet us. They helped us move down the road to Masitala village and into a one-room house near Uncle John.

This is where my father became a farmer and my childhood began.

Not LONG AFTER WE arrived, Uncle John acquired some additional land from Chief Wimbe, so he gave my father a one-acre plot about two kilometers from the house. There we could grow our own burley tobacco to sell, along with maize and other vegetables to eat. Maize is just another word for white corn, and by the end of this story, you won't believe how much you know about corn.

When we first arrived, Uncle John was busy planting his tobacco, which was the first item that needed my father's help. My father would wake up early before the first cock and go down to the grassy marshes in the valley, which we called *dambo*s. Because tobacco seeds require loads of water for them to break ground, many farmers plant nursery beds by the *dambo*s where they can easily water them daily. Each farmer has his own plot by the marsh—nothing official with papers or signatures, just a piece of ground you always know is yours. Not only is there water, but the soil in the *dambo* is deep black and full of nutrients that a little tobacco seedling requires to grow strong.

Making nursery beds is done just before the rainy season when the sun is the hottest. The work is hard and dirty, and my father quickly felt exhausted. During those first weeks, he'd dream of his stall at the trading center, how he used to just sit and chat with friends and customers, how he'd knock off at lunch for an hour to see his family, even take a quick nap before returning to work. It would have been easier to just tell his brother he'd made a mistake and return to Dowa, but my father buckled down and pressed on. He'd seen how much money Uncle John was earning, and he wanted the same for himself. Often he'd work so hard and late in the day that his brother would come looking for him, thinking he'd tripped and drowned in the *dambo*.

"Take a break, brother," he'd say. "Leave some for tomorrow. Reserve your strength, you'll need it."

"Just a bit longer," my father would say, his body covered in mud from head to toe.

WHEN UNCLE JOHN HAD visited Dowa and mentioned having a big place for my father, he wasn't talking about the living arrangements. With five people, our little house quickly became crowded.

After ten long hours of working in the sun, my father would come home and then start working on building our new house. Weekends were also spent this way. The bricks were fashioned out of grass and clay, which was pressed into a wooden mold about seventy-five centimeters long.

To get the clay, my father dug deep pits near the fields that swallowed his entire body. He'd scoop buckets of clay that weighed a hundred pounds, hoist them onto his shoulders, and climb out using steps he'd carved into the wall with a hoe. He'd then cart the pails two kilometers back to the house, dump them, then do it all over again.

After molding the bricks, my father spent days in the valley hacking the long-stemmed grasses to be used for roofing, then tied them into round bales. Sometimes John sent a few seasonal workers from his fields to help with the building, but my father did it mostly alone. After two

Me as a young boy in Masitala village, no doubt plotting some mischief to cause my mother grief.

months, we had a two-room house. Later, he'd say it was the hardest thing he'd ever done.

"Well done, brother," Uncle John said as he passed, joking with my father, who was about to collapse from exhaustion. "This is a good house. You know, every man needs a good house."

We lived in this house for three years until our growing clan became too big. Before long, there were five kids in our family, with me the only boy. By this time my father had earned enough on the farm to hire some men to construct two new buildings. The first one had a family room and master bedroom, plus a grain storage area. The other building, just across a narrow open corridor, had a kitchen, plus a separate bedroom for me and my sisters.

My bedroom became my fortress against the squabbling girls, a hideaway where I could be alone with my thoughts. I became a terrible daydreamer, partly because as I got older, the folktales of my childhood began to pale in comparison to the fantastic goings-on at the farm—things more real and incredible than any fiction my father could have imagined himself.

ONE OF THE SEASONAL workers Uncle John hired to help with planting and harvesting was named Mister Phiri, a man of near-heavenly strength. Uncle John didn't even use tractors to clear the land and trees. Instead he sent Phiri, who was so powerful he'd walk from tree to tree and rip them from the earth, as if they were weeds.

Everyone knew Phiri's secret was *mangolomera,* a form of magic that delivered superhuman strength. *Mangolomera* was the ultimate self-defense, a kind of vaccine against weakness. Only the strongest wizards in the district could administer this potion—a kind of paste made from the burned and ground bones of leopards and lions, and mixed with roots and herbs. The medicine was rubbed into small incisions made on each knuckle, usually by a magic razor. Once *mangolomera* was in your blood, it could never be reversed and was always gaining strength. Only the toughest men could manage this ever-growing power, or else quickly self-destruct.

Phiri was so strong that no person or animal could challenge him. Once while working in the fields, a black mamba snake slithered over his foot and prepared to strike. But Phiri wasn't afraid. He took a simple blade of grass and whipped the snake on the back, leaving it paralyzed. He then grabbed it by the head and snapped its spine. People said he carried another mamba in his pocket as a charm, and this snake was too afraid to bite.

But Phiri's power was so potent and always growing that it made him constantly want to battle. When this happened, my father had to intervene.

One afternoon I was playing in the yard when I heard a frightening noise coming from the fields, like the sound of twenty leopards roaring. I raced down to find Phiri nose to nose with another worker named James. Phiri was breathing heavily and ready to attack. His hands were in fists and the veins in his arms bulged like tree roots. When he opened his mouth to scream, the earth below our feet seemed to tremble in fright. Someone said Phiri had given James money to buy some items in Kasungu. But James wasn't educated and couldn't read or count, so the shopkeepers cheated him and kept their pay.

Before I knew it, Phiri began punching James. Phiri was short and thick, and James was tall and also very strong. The two traded blows back and forth, and for the moment, James was holding his own. But I knew it was only a matter of time before Phiri's *mangolomera* exploded and crushed poor James.

Around that time, my father also heard the commotion. Fearing for James's life, he rushed over to break up the fight. Although *mangolomera* never weakens, it can be neutralized for short periods of time using the green vines from a sweet potato plant. You know how Superman becomes weak at the sight of those shiny green crystals? The same is true for magic people and sweet potatoes, I don't know why.

Anyway, the second Phiri saw my father arrive, he shouted to him, "Mister Kamkwamba, *PLEASE* . . . some vines for my head! I don't want to kill this man!"

Seeing no vines nearby, my father instead ran over to Phiri and wrapped him up in his arms. Phiri kicked and screamed like a tethered tiger, but my father held on tight. He took him to our garden and pulled several long stems, then wrapped Phiri's head and elbows. Within seconds, Phiri's heart cooled down, and he collapsed from exhaustion. That day, seeing my father wrestle something as dangerous as *mangolomera* made me believe every story I'd been told about the Pope's awesome power.

The next morning, Phiri arrived for work looking and feeling okay. However, James reported being sick and had to miss the entire week. His hands and arms were so swollen he couldn't move, and his legs wouldn't even carry him. I'd watched James defend himself well, so this wasn't the result of Phiri's blows. Phiri's magic had been so strong it had simply rubbed off like poison.

PHIRI HAD A NEPHEW named Shabani who went around boasting that he was a real *sing'anga* who possessed *mangolomera*. Gilbert and I suspected he was just a lot of talk, but we were never completely sure. Shabani was a small boy like us and not that powerful, yet he boasted like a man with biceps the size of anthills. This made us wonder. Since Shabani never went to school, choosing instead to work the fields with his uncle, he was usually hanging around the house when I returned in the afternoons.

At the time, I was nine years old and not very strong. I wasn't the most athletic chap, either. Despite an incredible love for soccer, I wound up on the bench most every match. Bullies stalked and tortured me in the schoolyard. It was a time of crippling humiliation.

One day, after hearing another of my pathetic stories, Shabani took me aside.

"Every day you're complaining about these bullies, and I'm tired of hearing it," he said. "I can give you *mangolomera*. You can become the strongest boy in school. All the others will fear you."

Of course, possessing superpowers was my most frequent daydream. I'd imagine myself a Goliath on the soccer pitch, with legs like rocket launchers. With *mangolomera,* bullies would crumble at my touch and wet themselves from fright.

My father had always warned us against playing with magic. Now as Shabani stood there, smiling like a mongoose, I saw my father looking down at me, standing next to Jesus. I then felt my head shaking yes, and my mouth beginning to move.

"Okay," I said. "I'll take it."

"We'll do it in the blue gums behind Geoffrey's house," Shabani said. "Meet me there in one hour, and bring twenty tambala."

I arrived in the forest first and waited in the dark shadows, my mind racing with all the possibilities. Shabani then appeared through the trees. He held a black *jumbo* that sagged at the bottom, containing something heavy, something powerful.

"Are you ready?" he asked.

"Yah, I'm ready."

"Then sit down."

We sat down in the dirt and leaves and he opened the bag.

"We'll start with your left hand, cutting the knuckles and inserting the medicine into your veins. Then we'll do the right."

"Why the left hand first?"

"You're right-handed, man. Your right hand is the strongest. I'm giving you equal power, so your punches will be deadly from both sides."

"Oh."

He reached into the bag and pulled out a matchbox.

"In here are the blackened bones of the lion and leopard, along with other powerful roots and herbs."

He fished out a wad of paper that contained more black ash, which he began mixing with the other potion.

"These other materials are very scarce, found only on the bottom of the ocean."

"So how did you get them?" I asked.

"Look boy, I'm not just another person. I got them from the bottom of the ocean."

"Okay."

"I stayed there for three whole days. If I wanted to, I could take every person in your stupid village and put them into my scarf and sling them over my shoulder. Don't play around with me, *bambo*. If you want this kind of power, it will cost you lots of money. What I'm giving you is only a small taste."

I didn't even see him pull out the razor. It just suddenly appeared, and before I knew it, he'd grabbed my left hand and dug into my first knuckle.

"Ahh!" I screamed.

"Be still and don't cry!" he said. "If you cry it won't work."

"I'm not crying."

One by one, my knuckles began to swell with bright drops of blood that poured down my hand. Pinching the powder between his fingers, he rubbed it into the bloody wounds. It stung like hot coals. Once he finished with both hands, I exhaled with relief.

"See, I didn't cry," I said. "Do you still think it will work?"

"Oh yeah, it will work."

"When? When will I have power?'

He considered this for a second and said, "Give it three days to work its way through your veins. Once it's complete, you'll feel it."

"Three days."

"Yes, and whatever you do, don't eat okra or sweet potato leaves."

"I'll remember," I said.

"And lastly, tell no one," he added.

I walked out of the forest, looking down at my wounded, blackened hands, which by now had begun to swell. They looked tough. I imagined my arms swinging heavy at my sides like two thick hoe handles. A rush of confidence filled my lungs.

That evening, I hid in my room and spoke to no one. I went to bed feeling good. *I'm a big man now,* I thought, drifting off to sleep. *A big man.*

Three days was a long time to wait, but it worked with my plan. It was summer holiday, and the following morning I was supposed to travel to Dowa to spend time with my grandparents. Dowa was the perfect place to polish my powers before returning home a legend.

Well, three days crept by so slowly I thought I might die from boredom. I loved my grandparents dearly, but there wasn't much to do at their house. As I said, my grandmother was a tough lady who'd made her own bricks and was always putting me to work.

On the fourth day, I awoke and immediately felt different. Sitting up in bed, my arms felt light, yet hard as tree trunks. My hands were as solid as two stones. Heading outside, I took off running down the road to test my speed. Sure enough, I felt the wind in my face like never before.

That afternoon my uncle Mada invited me to watch a District League soccer game at the town pitch, and I went in hopes of testing my powers. The game was Dowa Medicals versus Agriculture, and as expected, the place was packed. As is our custom, the women looked after the children on one side of the field, while the men and boys huddled closely on the other, smoking cigarettes and shouting insults at the officials.

I had no interest in the game. I scanned the crowd until I saw a boy, perhaps my age, standing near the far corner of the pitch. He appeared to be alone, so I made my move. I cut through the crowd toward him, and as I walked past, I crushed his bare feet with my sandal. He let out a cry.

"Excuse me, you just stepped on my toes!" he shouted, hopping in pain.

I looked at him with two dead eyes.

"I said you stepped on my toes. It hurt."

"So?" I said.

"Well, it's rude, don't you think?"

"What are you going do about it?"

"What am I going to do?"

"You heard me. Why don't you do something, *kape*." A *kape* is a drooling idiot.

"Okay, fine," he said. "I'm going to beat you."

"That's what I was hoping you'd say."

We began dancing around in circles, and I wasted no time. I unleashed a flurry of punches so fast my arms became a blur in front of my eyes. I gave him lefts and rights and uppercuts for good measure, my two iron fists moving so quickly I couldn't even feel them smashing his face. Not wanting to kill the poor chap (I'd forgotten my potato vines), I finally backed away. But to my amazement, the boy was still standing. Not only was he standing, he was laughing!

Before I could release another deadly round, I felt a terrible pain in my right eye, then another, and another. Soon I was lying on the ground while his fists pounded my head and face, and his foot stomped my stomach.

By the time my uncle raced over and pulled him off me, I was crying and covered in dust.

"What are you doing, William?" my uncle shouted. "You know better than to fight. This boy is twice your size!"

Humiliated beyond anything I could imagine, I ran home to my grandparents and stayed inside until it was time to go home. And once there, I immediately found Shabani and confronted him.

"Your magic doesn't work! You promised me power, but I was beaten in Dowa!"

"Of course it works," he said, then thought for a second. "Listen, did you bathe the day I gave it to you?"

"Yes."

"Well, that's why. My medicine doesn't allow you to bathe."

"You never said that."

"Of course I did."

"But . . ."

As you can see, I was clearly cheated. My first and only experience with magic had left me with a sore eye and hands that throbbed from bad medicine. *With my luck,* I thought, *they'll probably become infected and fall off.* I began imagining myself a handless beggar in the market, unable to even use the bathroom. The fear of this occupied my mind for hours at a time. I'm telling you, it would be terrible!

CHAPTER THREE

IN JANUARY 1997, WHEN I was nine years OLD, OUR family experienced a sudden and tragic loss.

One afternoon while tending the tobacco with my father, Uncle John collapsed in the field. He'd been sick for several months but refused to see a doctor. That day, when my father helped him to the clinic near the trading center, they diagnosed him with tuberculosis and told him to go immediately to Kasungu Hospital. Uncle John's pickup wasn't running at the time, so my father ran to borrow a friend's car. Before he left, he placed his brother's bed mat under the cool shade of the acacia tree where he could rest. Uncle John's wife, Enifa, stayed by his side and kept him company, and soon, many others from the village joined them.

Not long after my father left, I heard a loud commotion under the tree, then panic. It was Enifa who began screaming first. I looked over and saw her push through the crowd, gasping for breath. Others around the tree soon began to wail and cry, holding their arms to heaven. I then felt a hand on my shoulder. I looked up and saw my mother, her face twisted as if she'd bitten something sour.

"Your uncle John is no more," she said. "He has passed."

It was then my father returned with the car and learned the tragic news about his brother. Several men had to hold his body up.

It was the first time I'd ever seen my parents suffer, and the sight of it frightened me more than any magic ever could. My uncle John was dead and his body lay under the acacia. I'd never seen a dead person, but I was too afraid to go look for fear it would never leave my mind. Soon I saw Geoffrey emerge from the crowd. He was crying and walking in circles as if he'd lost his direction. I didn't know how to behave, or what to say to him. I wanted to take my cousin and go away, down to the *dambo* where we could play and I could think. I didn't like the way I was suddenly feeling. You know, in our culture, when a loved one dies, you're expected to wail and cry to properly show your grief. I can't explain why, but I didn't feel like doing this. And after seeing everyone else, especially my father with his eyes red and face swollen from tears, I began to feel ashamed. So sitting there alone, I forced myself to cry, focusing on my dead uncle until I could feel the tears run hot down my face. Before they could dry, I went and joined my cousin to show my respect.

LATER THAT DAY, MY father's two brothers, Musaiwale and Socrates, arrived from Kasungu, along with other family and friends who'd heard the news. Members of the church also came to Uncle John's house and stayed all night and the following day. They pressed inside the two rooms and sang "This World Is Not My Home" while others quietly shuffled in and out to pay their respects. Uncle John's body lay on a grass mat on the floor covered with a brightly patterned *chitenje* cloth. The next morning a simple wooden coffin arrived from Kasungu and the body was delicately placed inside, yet I never gathered the courage to enter the house myself.

January is the rainy season when the air is thick and hot. As more and more people arrived that morning, the house became crowded and sticky, and the sound of people wailing became too much for Geoffrey to handle. At one point, he stepped out looking even more confused than before, and walked over to where I sat.

"Cousin, what next? What will happen?"

"I don't know," I said. What could I say?

For the rest of the day, Geoffrey would go inside, look at his father's body, then come back out and cry. He did this until it was time for the funeral to begin.

Chief Wimbe was out of town, so his messenger and bodyguard Mister Ngwata came to the house, along with other village headmen. For hours they sat under the acacia tree and discussed the funeral and what should happen with the family. When a powerful man dies, a lot of work needs to be done. In the event of a problem with the heir or transfer of property, it's the chief who must decide an outcome.

Finally everyone poured out of the house and gathered around the tree. Mister Ngwata stood and addressed them on behalf of Gilbert's father:

"We know this man has left behind some riches, and these treasures include his kids. We'd like to advise his brothers to take full control of these children. Make sure they finish their secondary education as they would have if their father had been alive. And in regards to the material wealth, we don't want to hear of troubles in the family as a result. If anyone here wants to help this family, help the children with clothing and school fees."

Another person stood up to speak. It was Mister Jonesi from Kasungu South, speaking on behalf of Geoffrey's mother's side of the family.

"This is a sad and tragic time even for our family," he said, holding his hat. "We're very concerned now. The deceased has left behind a wife, our beloved sister Enifa, and her four children. Our sister left our family long ago to join this village, so we ask the Kamkwamba side to please care for the kids and finish the job their dear father began. That's all."

My father and his brothers then lifted the coffin and placed it inside their friend Kachiluwe's truck. They jumped inside to hold the coffin in place as the truck rolled toward the graveyard. The crowd then followed on foot. The graveyard was located down the trail near Grandpa's village. It was just a small place under a grove of blue gums, with tall grass grown up around a few concrete headstones. My father's two sisters, Fannie and Edith, were also laid to rest there.

Several men dressed in gum boots were already waiting when every-one arrived. These were the *adzukulu,* or grave diggers, who are hired to do the job of digging and burying. In Malawi, graves are not just six-feet-deep open pits like those dug in Western countries. Instead, every grave has a hidden compartment at the bottom—usually a smaller cubbyhole carved into the side of the pit—where the coffin slides in. It's like having your own little bedroom in death. The purpose is to protect the deceased from the falling dirt, or really, to keep the family from seeing the falling dirt land on the coffin. For Uncle John's grave, the *adzukulu* had dug the compartment at the bottom center of the hole—a kind of hole within a hole.

Grunting, the *adzukulu* carefully lowered the coffin with ropes, into the smaller compartment. It was the exact size of the coffin. One of the gravediggers then jumped in and covered the hole with wooden planks and a reed mat. With its new floor, the open grave now appeared empty.

I watched all of this happen as if in a fevered dream, head throbbing, a dull buzzing deep in my mind, as if the pressing sun overhead had revealed to me its voice. Once the grave was finally filled and covered with grass, I joined the mourners back up the hill. It was the loneliest feeling I'd ever felt.

FOLLOWING UNCLE JOHN'S DEATH, things became more difficult all around. In addition to the sadness we all experienced, my father had to care for the business alone. It was the start of the growing season, and my father tended the crops through until harvest. He paid all the sea-sonal workers and settled all the accounts. Then, heeding the advice of the chiefs, he handed the entire business over to John's firstborn son, Jeremiah, who was twenty years old.

It's custom for the firstborn son to inherit everything from his fa-ther, but it doesn't always work that way. Often one of the brothers steps in and snatches control, leaving the family of the deceased at his mercy. This unfortunately happens all the time, and it's the number one grievance brought before the village chiefs.

Jeremiah lived at home with Geoffrey and their mother and often helped on the farm, but it was well agreed that he didn't like hard work. Although he was very smart, he'd never shown much interest in school and could often be found drinking in the boozing centers. My father felt terribly nervous about handing him the family business, but he wanted no trouble from chiefs or relatives.

"I don't want anyone saying I'm a thief," my father said. "If things go badly, I still did the right thing."

Of course, when Jeremiah heard he was being handed a family fortune, he was very surprised. He'd just assumed his father's brothers would never trust him.

"This is such a wonderful blessing," he told my father. "Thank you very much."

But as soon as Jeremiah took control, he spent most of the season's profits in the bars of Lilongwe and Kasungu. In November, when it came time to buy seed and fertilizer to plant new maize and tobacco, plus hire a new crew of workers, little of the money was left. As a result, the next crop was smaller. And when the tobacco was sold at auction, Jeremiah took the money and disappeared, returning only after most of it was gone.

Uncle John had also owned and operated two maize mills in nearby villages that made a substantial profit. In addition, he owned eight head of cattle. The mills and cattle were also given to Jeremiah, but the following year, Musaiwale, the oldest brother, forcefully took one mill and half the cows. Within two years' time, Jeremiah had lost both his maize mill and his cows.

As far as my father was concerned, his brother's business was gone. In farming, a man can lose everything so quickly. Given our custom, my father was forbidden to take back what he'd given away. Once you surrender control, you lose it forever. After the business collapsed, our family was left to survive on its own.

FARMING HAD ALSO BECOME a tougher business in Malawi, thanks to the policies of a new president. In 1994, three years before Uncle John's death, President Banda finally retired after losing the first elections he'd allowed to happen. Thirty years had been a long time in power, and the people were tired. Opposition against him had also grown ugly. Large crowds had gathered in the cities to protest his tyranny and harsh policies, and riots had erupted as a result. Before the election, Banda's thugs had even attempted to scare people into voting for him again. One day in the trading center, more than three hundred Gule Wamkulu appeared on the road carrying empty coffins, promising to fill them with anyone who didn't support the Life President.

But the opposition had still won, and unlike most African losers, Banda agreed to leave quietly and not start a war. He even accepted defeat before the final votes were tallied. He knew it was time. Since Banda had been born and raised in Kasungu, he returned to his home at the base of Mount Nguru ya Nawambe—formerly the Rock of the Edible Flies, where our great Chewa warriors had defeated the Ngoni—and lived out his final days. A big, fat former cabinet minister named Bakili Muluzi then became president, bringing with him his own brand of troubles.

Banda may have been a cruel dictator, but he did care deeply for farmers and the land. Our district is the most fertile in all Malawi, often called the "breadbasket" of the country, and Banda understood what was required to work the soil. He made sure that fertilizer was available to every farmer in the country who needed it. Seed was also cheap, allowing any Malawian to grow tobacco to sell. This meant that as long as it continued to rain, no family would go hungry.

On the other hand, Muluzi had been a wealthy businessman before entering politics and believed government had no business dealing in fertilizer and seed. He wanted to be different from Banda in every possible way, and this included stopping all subsidies and making the farmers fend for themselves. The free market allowed wealthy companies to flood the auction floors with mass-produced tobacco that drove the prices down and squeezed the small farmer. Soon, the value of our burley tobacco was so low that many farmers didn't bother growing it. My family managed to plant a few small plots, in addition to our normal maize fields. But without the help of seasonal workers, it was up to me and my cousins to help keep our farm running.

THE YEAR AFTER UNCLE John died, my uncle Socrates lost his job as a welder at Kasungu Flue-Cured Tobacco Authority when the estate closed. He and his family were forced to leave their quarters there and move back to our village, to a large shed near our house.

Uncle Socrates had seven daughters, which was good news for my sisters, but to me, their arrival didn't mean much one way or another. However, as we unloaded their things from the ten-ton lorry, I saw something leap from the truck bed.

Out of nowhere, a large dog appeared at my feet.

"Get away!" Socrates shouted, kicking the air above the dog's head. It yelped once and scampered off. Once at a safe distance, it sat down and stared at me.

"That's our dog, Khamba," he said. "I figured we'd bring him along to watch the chickens and goats here. That's what he did best at the estate. Maybe it'll remind him of home. We'll sure miss it there."

Khamba was the most unusual thing I'd ever seen: all white with large black spots across his head and body, as if someone had splattered him with a pail of paint. His eyes were brown and his nose was peppered with bright pink blotches. He looked exotic, like something from another land. Plus, he was big—much taller than the dogs around our village, but certainly just as skinny. In Malawi, dogs are kept only for security, and as a result, they aren't fed like their cousins in the West. Malawian dogs eat mice and table scraps, when there are any. In all my life, I'd never seen a fat dog.

Khamba sat there watching me, his long white tail fanning the dirt behind him. His long tongue hung out the side of his mouth, dripping saliva. As soon as Socrates went inside, Khamba came over and mounted my leg.

"Get away!" I shouted, making a swatting motion with my hands. The dog scurried against the house.

"Go chase some chickens, you stupid animal!"

His tongue came rolling out again, slobbering on the dirt.

The next morning when I awoke, I tripped over something as I stumbled out toward the latrine. There was Khamba, lying square in my doorway, ears perked and waiting.

"I thought I told you to leave me alone," I said, then realized what I was doing. I couldn't let anyone catch me talking to animals. They'd think I was mad.

Walking back from the toilet, I met Socrates coming out of our house with my father. He smiled and pointed at the dog now attached to my shadow.

"I see you found a friend," he said. "You know, the good Lord blessed me with seven children, but they're all girls. I think Khamba is happy to have found a pal."

"I'm no friend to a dog," I said.

Socrates laughed. "Tell that to him."

AFTER THAT, I GAVE up trying to get rid of Khamba. It was no use. And to be honest, he wasn't all that bad. Since I'd never had a dog of my own, it was nice having someone around, especially someone who didn't talk or tell me what to do. Khamba slept outside my door each night, and when it rained, he'd sneak into my mother's kitchen and curl himself in a corner. And without being told, he assumed his job as watchman over the goats and chickens, protecting them from the rare hyena or packs of mobile dogs that wandered wild and ate off the land. He also chased the goats through the compound, causing them to bleat and cry and kick up the dirt. When he did this, my mother would lean out of the kitchen and pitch one of her shoes at his head.

"Get that dog out of here!" she'd shout.

It was all a game to Khamba. He constantly tortured the chickens and guinea fowl, too, and even seemed amused when the mother hens flared their wings at him, hissing and giving chase.

But above all, what Khamba enjoyed most was hunting.

By this time, going hunting in the fields and *dambos* began to replace many of the games I used to play at home. I'd started by tagging along with my older cousins like Geoffrey and Charity, who also lived nearby.

Mostly we hunted birds. We hid in the tall grass by the *dambos*, which is so high during the dry season it can swallow a man whole. We'd wait until the afternoons when the birds came there to drink, then positioned a few sticks baited with *ulimbo,* a sticky sap that worked as a sort of glue. Once the birds stepped on the stick, they'd get caught and flap around, making all kinds of wild noises. Before they could break free, we'd jump out of the grass with our pangas, shouting:

"*Tonga!* I've got it!"

"*Tamanga!* Get it fast, so you don't scare off the others!"

"I'll cut its throat!"

"*No*—I want to pull off its head!"

We'd fight over who did the killing—usually taking turns cutting off the bird's head, or holding it between our fingers and—*thop*—pulling it like a tomato. We'd clean the insides, remove the feathers, and store them inside sugar bags we slung around our necks. Once home, we'd make a fire and roast the birds on the red embers. Fortunately, our parents never made Geoffrey and I share our hunting meals, and some nights during summer, we'd come home with eight birds and have quite a feast.

My family never had much money, and trapping birds was often our only way of getting meat, which we considered a luxury. The Chichewa language even has a word, *nkhuli,* which means "a great hunger for meat."

It wasn't easy to satisfy this hunger, and sometimes these missions proved to be treacherous. For one thing, the best *ulimbo* sap for trapping birds came from the *nkhaze* tree, which grew very thick with branches covered with thorns. One had to squeeze inside the *nkhaze* with his panga and cut the trunk, being careful not to get the sap in his eyes. If he did, he went blind.

One afternoon, Charity, Geoffrey, and I were out looking for *ulimbo* when we spotted the perfect *nkhaze* tree.

"I'll go!" said Charity. He was a kind of loud guy, who always wanted to be the leader. So we let him.

Charity climbed into the *nkhaze* tree with his knife, being careful of the sharp thorns all around. He reached up and sliced the trunk, then held a plastic sugar bag against the dripping wound. But just as he was doing this, a great gust of wind shook the entire tree, slinging the *ulimbo* into his eyes. Charity burst out of the bush, screaming, "I'm blind, I'm blind! Help me! It hurts!"

"What should we do?" I asked Geoffrey.

A man named Maxwell, who once worked for Uncle John, had taught us about the *nkhaze* tree and what to do if the sap ever got into our eyes.

Geoffrey turned to me. "You remember what Maxwell told us."

"Yah," I said. "What?"

"The only remedy is the milk from a mother."

"Oh, where are we going to find that?"

"Your house."

It was true, my mother had just recently given birth to my sister Mayless. Perhaps she could help. We guided Charity by the shirt and led him to my house. Once there, Geoffrey made our case to my mother, who happily agreed. She instructed Charity to kneel down and open his eyes. She took one breast from her shirt and leaned in close to his face.

"Hold still," she said, and squeezed a stream of white milk into his eyes.

It was hilarious. "*Eh* man," Geoffrey shouted. "Don't get any in your mouth!"

"This is your payment for satisfying *nkhuli*," I added, holding my ribs.

I never asked Charity how he felt about that incident, but I suppose it didn't matter. Within minutes, he was able to open his eyes and see. We all agreed that Maxwell must be some kind of wizard for knowing this secret. My mother told Charity, "For my services, I get all the birds you kill on your next hunt."

Charity agreed. The next day he brought four birds in a sugar sack and dropped them in the kitchen.

HUNTING WITH MY COUSINS had taught me the ways of the land: how to find the best spots in the tall grass and along the shimmering *dambo* pools, how to outwit the birds with a strong, smart trap, and the virtues of patience and silence when lying in wait. Any good hunter knows that patience is the key to success, and Khamba seemed to understand this as if he'd been hunting his entire life.

Our first outings began with the start of the rainy season, when the showers are heavy all morning and replaced in the afternoon by a swelter-ing, pasty air. When the land is wet and filled with puddles, the *dambos* don't attract as many birds. This is when we hunters rely on the *chikhwapu*, a giant deadly whip—or a kind of slingshot trap without the stone.

After the rains stopped one morning, Khamba and I set out to make our trap. I carried a sack on the end of my hoe made from a *mpango*—a kind of long, brightly colored scarf used by women to hold everything from their hair to babies on their backs. The sack contained a long bicycle tube, a broken bicycle spoke, a short section of steel wire I'd clipped off my mother's clothesline, a handful of maize chaff we called *gaga*, and four heavy bricks. As always, I also carried the two hunting knives I'd made myself.

The first was a Rambo-style commando knife I'd made from thick iron sheets. First, I'd traced a fierce-looking pattern on the metal with a pencil. Using a nail and heavy wrench, I poked holes all along the lines, perforat-ing the metal so it popped out with a good pounding. I then ground the metal against a flat rock to smooth the edges and produce a sharp blade. For a handle, I wrapped the bottom of the blade in enough plastic *jumbo*s to get a full, even grip. Then I melted the handle over a fire.

My second knife was more like a stabbing tool made from a large nail I'd pounded flat with the wrench and ground to a sharp edge. I'd fashioned its handle in the same way as the first. I kept both knives tucked snugly in the waistband of my trousers.

Packing my gear, I set off with Khamba down the trail behind Geof-

The view of the Dowa Highlands from my home. The mountains lie just beyond the maize rows and blue gum forest where Khamba and I would hunt.

frey's house that led to the graveyard, down into the blue gums where the trees were taller and provided good shade. The hills of the Dowa Highlands—which separated us from the lake—rose beautifully before me, capped in gray, dripping thunderheads. A new storm was on its way, so we had to work quickly.

I found a good spot off the main trail, near a tall blue gum that would cast a long shadow once the sun broke through the haze. Using my hoe, I cleared away the grass and vines until the red mud was exposed—a surface of about four feet in diameter. Taking my knife, I sawed off two thick branches from the blue gum and stripped their bark, then whittled both to sharp points. I pushed the poles into the moist soil about two feet apart, then pulled them to test their firmness. They held.

I cut the bicycle tube into two thin strips and attached both pieces to the section of steel wire. I then tied the rubber strips to the blue gum poles. When finished, it resembled a giant slingshot with a thick steel center. This was the kill bit.

Stripping bark off several nearby trees and lashing it together, I fashioned a long rope about fifteen feet long. I then cut a small, eight-inch section off it and attached it to the steel bit. I tied a short stick to the other end, making the knot fat and round. Gripping the stick like a handle, I pulled back the rubber bands as far as they'd stretch, then wedged the handle between two posts—a second stick and the bike spoke—using the fat, round knot to hold it in place. The long rope then led back into the trees and acted as the trigger. Once it was set, I stacked the four bricks several inches in front of the trap, then sprinkled the maize chaff in the middle. This was the kill zone.

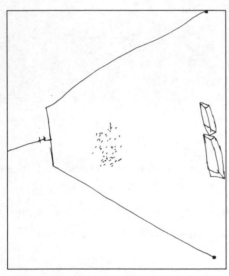

The chikhwapu *trap used to kill birds during the rainy season. The birds smashed into the bricks and died. Then I ate them.*

When the birds landed to eat the chaff, I'd pull the rope and release the sling, slamming the birds into the wall of bricks.

"Let's hunt," I said, and Khamba followed me into the trees.

We hid behind a small *thombozi* tree that allowed me to see clearly without being spotted. As soon as we got there, Khamba lay down beside me and stared keenly ahead. He never moved, never barked. After about half an hour, a small flock of four birds swooped over and spotted the bait. They fluttered down and began pecking at the dirt. My heart began to race. Khamba's ears perked up and his mouth began to quiver. I was about to release the rope when I saw a fifth bird land just behind the others. It was giant, with a fat gray chest and yellow feathers.

Come on, I thought, *a little more to the right. That's it, come on.*

After a few long seconds, the fat bird nudged his way into the group and started to feed. Once they were square in the kill zone, I pulled the rope.

WHOO-POP!

The birds disappeared in a cloud of feathers and chaff.

"*Tonga!*" I shouted, and Khamba and I dashed out of our blind.

Four birds lay dead against the bricks, while a fifth had managed to fly away. The large bird was still flapping against the mud, so I picked it up before it revived. Its body was warm and soft in my hands. I could feel its tiny heart fluttering against my palm. I pinched its head between my two fingers and twisted its neck.

I picked up the others and dusted off the mud. Normally I'd carry a sugar bag, but today I'd forgotten. I stuffed the limp birds into my pockets.

Once the trap was reset, I waited for another half hour, then finally gave up.

"It's time to eat," I said.

Khamba and I then set off for *mphala*.

MPHALA MEANS "A HOME for unmarried boys," which is exactly where my cousin Charity lived. It was more like a clubhouse, situated on our property just across from Geoffrey's house. James, the seasonal worker who'd fought Phiri, had once lived there. But after he'd been laid off, it remained empty. Charity had taken over the house with his friend Mizeck, a big fat guy who'd dropped out of school and now worked as a trader. Although they both still lived with their parents—Charity's house was near Gilbert's in the blue gum grove—they slept at the clubhouse at night.

In the corner, someone had built a bed from blue gum poles and maize sacks stuffed with grass. Dirty clothes were strewn everywhere, along with mango peels and groundnut shells and other strange pieces of rubbish. One wall featured a poster of the soccer club MTL Wanderers—otherwise known as the Nomads—which were my favorite team in the Malawi Super League, and possibly the whole world. A poster of their chief rivals, Big Bullets, adorned the opposite wall, and I can't tell you how much I hated Big Bullets. A fireplace sat in the corner—just a large shallow pot with holes poked in the sides for ventilation and filled with charred maize piths and wood. A small window above ventilated the smoke, but not very well. It also let in the room's only light, a thin beam of sunshine that was polluted with hanging dust. The air stank like dirty feet. To me, it was the greatest place in the world.

Because I was young and annoying, I was mostly forbidden from entering the clubhouse, unless, of course, I earned my entry. A few times I was allowed in after helping steal mangoes. Charity would make me wear a *mpango* sack around my neck and sneak into the neighbor's compound. With my knife in my teeth, I'd climb the trees and quietly snip the mangoes and drop them into the sack. I'd take them back to *mphala* and they'd let me inside. It was like paying dues.

Once inside, the conversation was lurid and often confusing for my eleven-year-old mind. Much of the talk was about girls, and I was lucky if they forgot I was there. One time, Mizeck stopped midway through a story about a certain girl he'd seen in town and said to Charity, "We should take care, we have a child among us. This boy can't handle such stories."

I started pleading. "I'm not a child. Come on guys, carry on. I'm a big man. I know some things about girls."

"Oh yeah, and what do you know?"

"I know . . . I know what you know."

As Khamba and I walked home from the hunt, I knew I'd earned enough loot to gain myself an entry. As I got close, I heard Charity and Mizeck inside. I knocked and Charity swung open the door.

"What?"

"Guys, I got four birds just now! They're here in my pockets. Can I come in?"

Mizeck appeared at the door. "What do you have for us?"

"Four birds."

He smiled. "This is the type of man we need here at *mphala*. Good job."

"We'll make a fire," said Charity.

I walked inside beaming. Khamba followed.

"Get that stupid dog out of here," shouted Mizeck. "He's going to think he lives here or something. Dogs don't belong inside, don't you know this? I bet you even talk to that thing."

"Khamba," I screamed, "get outside!"

I reared back my leg, and he scurried out the door, then looked at me confused.

"Just wait," I whispered.

I began cleaning my birds, plucking off the feathers and shaking them from my fingers into a pail. I popped the heads off and scooped out the entrails. When I opened the door, Khamba was waiting. This was his hunting treat, a reward more treasured than life itself. I tossed each head into the air, and Khamba leaped up and grabbed them. One crunch and they were gone. The entrails were slurped in a gulp.

Back inside, Charity and Mizeck already had the birds laid over the coals. The sizzling meat smelled delicious.

"Guys," I said, "I'm really starting to salivate!"

"Be quiet."

Once they finished cooking my birds, they even allowed me to eat one. But as soon as I was no longer useful to them, the inevitable happened.

"Hey boy," said Mizeck, "don't I hear your mother calling?"

"What? I don't hear anything."

"He's right," said Charity. "That's definitely your mother."

My marching orders had been given. Without protest, I holstered my knife back into my waistband, called my dog, and together we returned home to a houseful of girls.

CHAPTER FOUR

THE YEAR I TURNED thirteen marked the beginning of a new century, and gradually, I noticed a change happening in myself. I started growing up.

I stopped hunting as much and started hanging out more in the trading center, socializing and meeting new people. Gilbert was usually with me, along with Geoffrey and a few others. We'd go there and play endless rounds of *bawo,* a game that's very popular in Malawi and East Africa. *Bawo* is a mancala game played with marbles or seeds on a long wooden board lined with holes. Each player had two rows of eight holes each. The object is to capture your opponent's front row of marbles and prohibit him from moving.

Bawo requires a lot of strategy and quick thinking. I'll admit, I was pretty good at this game and would often beat the other boys at the trading center, a small revenge since most of them had benched me in soccer when we were younger. If I never got *mangolomera,* at least I had *bawo.*

Each time I left for the trading center to see my friends, Khamba would perk up and try to follow me. He missed our trips together, but I forbade him to tag along. People would think I was backwards for walking with a dog. One time Khamba followed me to the trading center without my realizing he was there. When I got to the fig tree near the barbershop where we played *bawo,* someone pointed and laughed.

"Why do you need this dog behind you?" they said. "I don't see any rabbits or birds around. Are you going hunting in the market?"

The other boys started laughing too. It was embarrassing. After that, whenever Khamba tried to follow, I had to get mean.

I cursed and shouted, but of course, he never listened. After a few meters, I had to pick up a small stone and hurl it toward his head.

"Now leave me alone!"

After a few times, he got the message. He'd still come to the trading center on his own, usually during July mating season, when the female dogs were in heat and roaming the villages. He'd see me and gallop over, wagging his long tail. I'd always stop him short.

"Get!" I'd shout, kicking the dust to scare him before anyone saw me.

Also, as I got older, the day-to-day fate of the MTL Nomads no longer determined my moods and emotions. Throughout my life, the Nomads had been more than men. I listened to every game on Radio One and imagined them as giants. When the Nomads lost—especially to Big Bullets—I became so upset I couldn't even eat supper, not even if my mother served chicken, and I loved chicken. This following had become an obsession. During a game that year with Big Bullets, my heart started beating so quickly I was convinced I was dying (I think they're called anxiety attacks). I thought, *What am I doing to myself? Soccer is too stressful for my health.* After that, I sort of stopped following the game altogether.

AROUND THIS SAME TIME, Geoffrey and I started taking apart some old broken radios to see what was inside, and we began figuring out how they worked so we could fix them.

In Malawi and most parts of Africa that don't have electricity for television, the radio is our only connection to the world outside the village. In most places you go, whether it's the deepest bush, or the busy streets of the city, you'll see people listening to small, handheld radios. You'll hear Malawian reggae or American rhythm and blues from Radio Two in Blantyre, or Chichewa gospel choirs and church sermons from Lilongwe.

Ever since the Malawi Broadcasting Corporation began, around the time of independence, Malawians have thought of their radios like members of their families. My father talked about the early days of MBC and hearing Dolly Parton and Kenny Rogers from America and the wonderful sounds of Robert Fumulani. Back then, agriculture programs were very popular, and my father remembers President Banda—Farmer Number One—reminding everyone to clear fields, dig ridges, and plant before the rains, saying that doing so would make Malawians happy and successful. He also reminded people to apply manure! And for me growing up, I'll always remember listening to the Sunday sermons of Shadreck Wame from the Church of Central Africa Presbyterian in Lilongwe, followed by the Sunday Top Twenty.

Unfortunately, until a few years ago, there were only two radio stations—Radio One and Radio Two—that were both operated by the government. This greatly reduced our window into the outside world.

From the first time I heard the sounds coming from the radio, I wanted to know what was going on inside. I'd stare at the exposed circuit boards and wonder what all those wires did, why they were different colors, and where they all went. How did these wires and bits of plastic make it possible for a DJ in Blantyre to be speaking here in my home? How could music be playing on one end of the dial while the preacher spoke on the other? Who'd arranged them this way, and how did this person learn such wonderful knowledge?

Through nothing more than trial and error, we discovered that the white noise was caused by the integrated circuit board, the biggest piece, which contains all the wires and bits of plastic. Connected to the integrated circuit are little things that look like beans. These are transistors, and they control the power that moves through the radio into the speakers. Geoffrey and I learned this by disconnecting one transistor and hearing the volume greatly reduce. We didn't own a proper soldering iron, so to perform repairs on the circuit boards, we heated a thick wire over the kitchen fire until it became red hot, then used it to fuse the metal joints together.

We also learned how the radio picks up each band, such as FM, AM, and shortwave. To receive AM, the radio has an internal antenna because its waves are long, but in order to receive FM, the antenna must be outside and reach into the air to catch the smaller, more narrow waves. Just like light, if FM waves hit something like a tall tree or building, they're blocked.

Since we learned everything through experimenting, a great many radios were sacrificed for our knowledge. I think we had one radio from each aunt and uncle and neighbor, all in a giant tangle of wires we kept in a box in Geoffrey's room. But after we learned from our mistakes, people began bringing us their broken radios and asking us to fix them. Soon we had our own little business.

We operated out of Geoffrey's small bedroom, located just behind his mother's house. There we waited for customers, the floor below us strewn with heaps of wires, circuit boards, motors, shattered radio casings, and unidentifiable bits of metal and plastic that appeared along the way. Our usual exchange with clientele went something like this:

"*Odi, odi,*" someone said, standing at the door. It was an old man from the next village, hiding his radio in his armpit like a chicken.

"Come in," I said.

"I heard someone here fixes radios?"

"Yes, that would be me and my colleague, Mister Geoffrey. What's the problem?"

"But you're so young. How could this be?"

"You mustn't doubt us. Tell me the problem."

"I can't find the station. It won't listen."

"Let me see . . . hmmm . . . yes, I think we can manage. You'll have it before supper."

"Make it before six! It's Saturday, and I have my theater dramas."

"Sure, sure."

If we were going to determine what was broken in the radios, we needed a power source. With no electricity, this meant batteries. But batteries were expensive, and Geoffrey and I didn't earn enough from our repairs to keep buying them. Instead, we'd walk to the trading center and look for used cells that had been tossed in the waste bins. We'd collect maybe five or six, along with an empty carton of Shake Shake booze. Even after all these years, I was still finding uses for these stinking cartons.

First we'd test the battery to see if any juice was left in it. We'd attach two wires to the positive and negative ends and connect them to a torch bulb. The brighter the bulb, the stronger the battery. Next we'd flatten the Shake Shake carton and roll it into a tube, then stack the batteries inside, making sure the positives and negatives faced in the same direction. Then we'd run wires from each end of the stack to the positive and negative heads inside the radio, where the batteries normally go. Together, this stack of dead batteries usually contained enough juice to power a radio.

Of course, success also depended on the brand of batteries and what they'd been used for. Handheld radios use very little power so they often drain a battery of its last drop, while cassette players require such high voltage that a battery can't maintain it and fails, even though it leaves a bit of juice. The worst batteries (but unfortunately the most common) were the Chinese-made Tiger Heads that died after several hours in any player. That's why we got so excited when we came across a beloved Malawian Sun battery, which were by far the strongest and powered our radios like none other.

"Mister Geoffrey, how lucky we are to have found a good Malawian Sun."

"You're right, mister man, we'll get plenty of time out of this one."

Often while we fixed our radios, people would approach us and say, "Look at the little scientists! Keep it up, boys, and one day you'll have a good job."

I'd become very interested in how things worked, yet never thought of this as science. In addition to radios, I'd also become fascinated by how cars worked, especially how petrol operated an engine. *How does this happen?* I thought. *Well, that's easy to find out—just ask someone with a car.* I stopped the truckers in the trading center and asked them, "What makes this truck move? How does your engine work?" But no one could tell me. They'd just smile and shake their heads. Really, how can you drive a truck and not know how it works?

Even my father, who I assumed knew everything, said: "The fuel burns and releases fire and . . . well, I'm not really sure. "

Compact disc players were just getting popular in the trading center, and these fascinated me even more. I'd watch people insert this shiny plate into their radios and hear music.

"How did they put the sound on that?" I'd ask.

"Who cares?" most people would answer.

Although the people in the trading center were content to simply enjoy these things without explanation, these questions constantly filled my mind. If solving such mysteries was the job of a scientist, then a scientist is exactly what I wanted to become.

At the time, I attended Wimbe Primary School, located a kilometer down the wooded path past Gilbert's house. The following year I'd sit for my Leaving Certificate Examination. If I passed, I'd advance to secondary school, where I heard students had more extended lessons in science and were even assigned experiments to conduct.

To me, being a scientist was worlds better than farming, which by then had started taking up a large part of my time. My father was still growing a bit of tobacco to sell at auction, but our main crop was always maize—or *chimanga*—which fed our family year-round. Most Malawians were sustenance farmers who depended on their maize plots to survive. If you weren't able to get food from anywhere else, at least you had grain in storage and your family could eat. Even people who lived in the city relied on their brother or nephew in the village to tend a plot of maize for them. Everyone needed this during the growing season when grain prices were high in the markets.

In Malawi, we eat maize with every meal, and most families serve this in the form of a doughlike porridge called *nsima* (pronounced like "seema"). *Nsima* is made by adding maize flour to hot (but not boiling) water until it becomes too thick to stir, then scooping it into cakes about the size of American hamburger patties. Tearing off a piece, you roll the *nsima* into a ball in your palm, then use it to scoop up your relish—usually beans or leafy greens, such as mustards, rape, or pumpkin leaves, whatever happens to be in season. If your family is fortunate, maybe you also have some goat or chicken. My favorite is dried fish with tomatoes!

Everyone from the fat politicians to the dogs and cats depends on *nsima* to live. Each night after our supper, Khamba would be waiting by his food bowl to get his delicious helping. Most of the time he didn't even chew his portion, just inhaled it whole.

"How can you even enjoy it?" I'd ask.

Nsima isn't just an important part of our diet—our bodies depend on it the same way fish need water. If a foreigner invites a Malawian to supper and serves him plates of steak and pasta and chocolate cake for dessert, but no *nsima,* he'll go home and tell his brothers and sisters, "There was no food there, only steak and pasta. I hope tonight I can sleep."

Farming *chimanga* is a family activity that requires the help of every man, woman, and child who is old enough to work. Young girls usually help a bit with planting, weeding, and harvesting, but mostly they assist their mothers with the many chores around the home, such as fetching water, cooking, cleaning, and taking care of little ones. In Malawi, a woman's contribution is often overlooked. By the time I was twelve years old, I had five sisters and no brothers, which meant I was the one who helped my father in the fields.

Work begins in July to clear the land from the previous harvest in May. We had to collect the dried maize stalks, stack them into heaps, called *chikuse,* and line them up in rows. Once all the *chikuse* were properly arranged I'd set them on fire and wait. Grasshoppers make their homes in these stacks, and once the stalks start burning, the grasshoppers fly out by the hundreds and are easy to catch. I'd throw them in sugar bags, then take them home to roast over the fire with salt. I'm telling you, I can eat huge amounts of *nsima* with crunchy grasshoppers. Of course, I wasn't supposed to be hunting grasshoppers while working, but we have a saying: "When you go to see the lake, you also see the hippos."

We spend much of August through November digging new dirt rows, or ridges. Taking a hoe, I'd hack apart the two existing ridges and form a new one in the center. This was our way of rotating the soil. Since this was our dry season, the soil was hard and required my full strength to break it apart, leaving large blisters on my hands. Not only that, but the hard soil left lumps that had to be crushed using the handle of the hoe, which added to the time this took. Soft soil allowed the seeds to pass through with no problem. Some farmers were lazy and left these lumps, and as a result, their yields were smaller.

Making ridges was always done while the sun was boiling in the sky. It was so hot I'd work in the mornings before school, then return in the evenings before dark. If the moon was full and bright, I'd wake up at 4:00 A.M. before the cock even crowed. In the still-dark of morning, I stumbled into the latrine with my torch, trying not to notice the spiders on the ceiling with their black-and-white legs and giant hairy bodies, or the cockroaches flickering their antennae in the light, as if telling me, *This is our time to play, you should be in bed!* In that quiet time of morning, you could even hear the termites eating the walls, which always sounded like someone walking outside in the grass. Back at the house, I'd draw a bucket of water from the shallow well behind our house and wash my face. (It wasn't potable.) By then, my mother would be up preparing a bowl of maize porridge—known as *phala*. After quickly devouring this, I headed down the trail with my hoe dragging behind.

"Make sure you watch where you throw that hoe in the dark," my father would call out. "I don't want you cutting off your foot!"

"For sure."

Chopping your foot is a common accident during the clearing and planting season. You'll often see children with plastic sugar bags or news-papers wrapped around their feet with twine—some form of makeshift bandage—to keep away the flies and soil. This doesn't work, however, since you're back out in the fields the next morning. Cuts don't always heal properly during this time, and every Malawian who ever grew up in a village has his own trail of scars to prove it.

Even with the moon, the road was dark and filled with shadows. I walked quickly and focused on each step, trying not to imagine the Gule Wamkulu watching from the trees, or bald men on witch planes flying overhead. No matter where I am or how old I get, these things will always scare me at 4:00 A.M. One morning while walking, a hyena cried from the bush—*ooooo-we*—practically causing me to jump out of my trousers. I've never run so fast in all my life.

Normally the rains arrive by the first week of December and continue through March. The first sign of rain is our signal to begin planting. It's like the starting pistol in the great race against God—the moment he says "GO!" When the first rains fall, you must be ready.

One person will take the hoe and move quickly down the ridges, making small gashes in the soil for planting stations while another follows and drops three seeds inside, then covers them with soil and a lot of good wishes. The fields in December are thick and muddy and stick like cakes against your feet.

After a few days of rain, the seedlings will push through the soil and unfold their tiny leaves. Two weeks later, if the rain is still good, we then carefully apply the first round of fertilizer, because each seedling requires love and attention like any living thing if it's going to grow up strong.

Making planting stations was my favorite job because I didn't have to wait on anyone in front, allowing me to knock off early and go fry maize—which in December is a rare pleasure.

From May through September, you're still benefiting from the harvest. The maize is abundant, and every meal is grand. That is our winter in the Southern Hemisphere, and on those cold evenings, we'll all gather round a fire and roast maize in a flat pan, laughing, telling stories, and humming good songs. But come the December planting season, most families are beginning to run low on maize, so any opportunity to sit with cousins and friends and fry maize is a treat. The smell of it roasting is heavenly and makes us all very happy. Also during this time of scarcity, every meal of *nsima* is eaten with a heavy heart.

December is when people must buy fertilizer and seed, which is expensive and exhausts their savings. Most people will save enough to buy a chicken and some rice on Christmas and New Year's, but afterward, there's nothing. Come January, most people are forced to tighten their belts and wait until harvest. Outside, it's always raining morning and night, and even the birds have nothing to eat because everything's too busy growing. It's a time of heat and mud and waiting. We call this period "the hungry season." In the countryside, people are working the hardest they work all year to prepare their fields, but doing so with the least amount of food. Understandably, they grow thin, slow, and weak. Children sometimes die. The hungry season has always been with us, as predictable as the cock and the morning sun.

But if all goes well, the steady rains of December and January will have allowed the seedlings to grow, and by now they should reach my father's knees. The small ears of maize then begin to form, and after some weeks, they reveal their blossoms—a cluster of silky hair and a tall tassel flower. By February, the stalk is thick and strong and as high as my father's chest. By harvest time in May, especially with fertilizer, the stalks can be well over his head. The maize is then left to dry on the vine, then pulled and plucked. The grain is stored in fifty-kilogram bags in a small storage room next to where my parents sleep. In a good year, the bags will rise to the ceiling and spill into the corridor.

THIS IS HOW WE normally planted and harvested each year, but in December 2000, everything went wrong. The rains were late, and didn't begin until the last week of December. The first showers gave the seedlings confidence to finally push through the soil, so farmers applied their fertilizer and hoped for the best. But the rains that followed were much too heavy, falling day and night for a whole week. Great floods swept across the country, carrying away homes and livestock, along with the seedlings that had just begun to grow. Fortunately there was no flooding in our district, but the rains still washed away the fertilizer and any hopes of high yields.

But like us, many farmers hadn't been able to purchase fertilizer anyway. As a result of the new president's policies, a bag of NPK fertilizer (consisting of nitrogen, phosphorus, and potassium) now costs three thousand kwacha. It was far too expensive to buy once, let alone twice if the rains washed away the first round. After the floods, the president went on Radio One and vowed to assist every farmer with "starter packs" that contained two kilos of maize seed and five kilos of fertilizer. The starter packs had been implemented back in 1998 and 1999 and given to every farmer in Malawi who needed one. Those years the rains had been good, and with the additional seed and fertilizer from the packs, the yields had been high. But due to pressure from international donors, the program had been slashed to about one million farmers. So it was nice to hear the president promising to broaden this assistance.

But a month passed and nothing happened. A government list then appeared in the trading center of those who'd receive packs. My father's name was missing, along with that of many others. But this hardly mattered. By then, it had stopped raining altogether.

After the floods, the rains simply vanished and a period of drought cursed the land. Each day the sun rose hot in the sky and showed no mercy on the young seedlings that had survived. By February, the stalks were wilted and hunched toward the ground like an old woman sweeping the dirt. A bit of rain in March saved us from total disaster and allowed the stalks to mature, but just barely. By May, the sun had burned half the crop. The stalks that survived were so stunted they only reached my father's chest.

One afternoon I walked out into the fields with my father and we looked out across this destruction. The maize leaves looked like onions, brown and brittle and ready to crumble to the touch. We were thinking the same thing, but I said it first.

"What will happen to us next year, Papa?"

He let out a sigh. "I don't know. But at least we're not alone. It's happening to everyone."

My father was right. In many parts of the country, the yields were far smaller. Droughts hurt the smaller villages the most since the tiny farms had to feed large families year-round. The slightest problem in weather, fertilizer, or seed productivity could tip these families off the edge into hunger. That year, the drought would be felt for several more seasons.

As for our own farm, we managed to fill only five sacks with grain that barely filled one corner of the storage room. One night before bed, I saw a lamp flickering inside and found my father there alone, staring at the sacks as if he'd just asked them a question. Whatever they told him, I'd find out soon enough.

CHAPTER FIVE

URING THIS TIME OF trouble, I discovered the bicycle dynamo.

I'd always seen them on bicycles, the way they were attached to the wheel like a tiny metal bottle, but I'd never paid them much attention. But one evening, my father's friend rode up to our house on a bicycle with a lamp powered by a dynamo. As soon as he hopped off the bike, the light switched off.

"What made the lamp go off?" I asked. I hadn't seen him turn a switch.

"The dynamo," he said. "I stopped pedaling."

Once he went inside to see my father, I jumped on his bike to try it myself, to see if I could make the lamp work. Sure enough, after a few meters of pedaling, the light came on. I got off, flipped the bike over, and traced the wires from the lamp all the way down to the rear tire, where the dynamo was attached. The dynamo had its own metal wheel that pressed against the rubber. Turning the pedal with my hand, the tire spun round and also spun the wheel of the dynamo. Then the light came on.

I couldn't get this out of my head. How did spinning a wheel create light? Soon I was stopping everyone with a dynamo and asking them how it worked.

"Why does the light come on when you pedal?" I'd ask.

"The dynamo is rotating, that's why."

"I know it's rotating, but why does it work? What is the secret?"

"I don't know."

"Can I play with it?"

"Help yourself."

I spun the wheel and watched the light. One day while playing with my father's friend's bike, I noticed the wires had come loose from the bulb. With the wheel spinning, I accidentally touched the ends to the metal handlebar and saw a spark. This gave me my next idea.

One afternoon Geoffrey and I borrowed that same bike, flipped it upside down, and connected the wires to the positive and negative heads on a radio, where the batteries would normally go. When we cranked the pedal, nothing happened. I then attached the wires directly to the base of the dynamo's bulb. This time when I pedaled, the light flickered on. Taking the batteries I'd removed from the radio, I stacked them together and ran a separate wire from the batteries to the bulb. Again, the light worked.

"Mister Geoffrey, my experiment shows that the dynamo and the bulb are both working properly," I said. "So why won't the radio play?"

"I don't know," he said. "Try connecting them there."

He was pointing toward a socket on the radio labeled "AC," and when I shoved the wires inside, the radio came to life. We shouted with excitement. As I pedaled the bicycle, I could hear the great Billy Kaunda playing his happy music on Radio Two, and that made Geoffrey start to dance.

"Keep pedaling," he said. "That's it, just keep pedaling"

"Hey, I want to dance, too."

"You'll have to wait your turn."

Without realizing it, I'd just discovered the difference between alternating and direct current. Of course, I wouldn't know what this meant until much later.

After a few minutes of pedaling this upside-down bike by hand, my arm grew tired and the radio slowly died. So I began thinking, *What can do the pedaling for us so Geoffrey and I can dance?*

The dynamo had given me a small taste of electricity, and that made me want to figure out how to create my own. Only 2 percent of Malawians have electricity, and this is a huge problem. Having no electricity meant no lights, which meant I could never do anything at night, such as study or finish my radio repairs, much less see the roaches, mice, and spiders that crawled the walls and floors in the dark. Once the sun goes down, and if there's no moon, everyone stops what they're doing, brushes their teeth, and just goes to sleep. Not at 10:00 P.M., or even nine o'clock—but seven in the evening! Who goes to bed at seven in the evening? Well, I can tell you, most of Africa.

Like most people, my family used kerosene lamps to find our way at night. These lamps were nothing more than a Nido powdered milk can with a cloth wick, filled with fuel and bent closed at the top. The fuel was very expensive, and the only place to buy it for a good price was at the petrol station in Mtunthama seven kilometers away. The lamps produced thick black smoke that burned your eyes and made you cough. Of course, you can buy hurricane lamps made with glass and a top that prevents the smoke from coming out, but most people can't afford these.

Our country's power is supplied by the government through the Electricity Supply Corporation of Malawi (ESCOM), which produces electricity using turbines on the Shire River in the south—another object of my fascination that I'll describe later.

If you have money and a lot of patience, you can ask ESCOM to bring electricity to your home by wire. You have to catch a pickup to Kasungu, then a minibus one hundred kilometers to the capital Lilongwe, where you'll find the ESCOM office in Magetsi House. There you pay someone thousands of kwacha, submit an application, and draw them a map of your home so they can find it. And if you're lucky, maybe your application will get approved and the workers will find your home and install a pole and wires—all of which you must pay for. When this power is finally on, you're happy staying up until ten o'clock dancing to your radio, but only until the government issues power cuts, and those happen every week, usually at night, just after dark. All that money and trouble—it was almost easier just going to bed at seven.

Another thing that contributes to our energy problem is deforestation. As my grandpa told me, the country was once covered in forests, with so many trees the trail grew dark at noon. But over the years, the big tobacco estates had taken much of the wood, using it to flue-cure the leaves before bringing them to auction. Local tobacco farmers used more wood to build shelters for drying the leaves, but these structures never lasted more than a season because of the termites. The rest of the wood got used by everyone else for cooking since we had no electricity. The problem got so bad near Wimbe that it's not uncommon for someone to travel fifteen kilometers by bike just to find a handful of wood. And how long does a handful of wood last?

Few people realize this, but cutting down the trees is one of the things that keeps us Malawians poor. Without the trees, the rains turn to floods and wash away the soil and its minerals. The soil—along with loads of garbage—runs into the Shire River, clogging up the dams with silt and trash and shutting down the turbine. Then the power plant has to stop all operations and dredge the river, which in turn causes power cuts. And because this process is so expensive, the power company has to charge extra for electricity, making it even more difficult to afford. So with no crops to sell because of drought and floods, and with no electricity because of clogged rivers and high prices, many people feed their families by cutting down trees for firewood or selling it as charcoal. It's like that.

One of these government power lines that serviced the nearby tobacco estates also connected to Gilbert's house. Since his father was Chief Wimbe, they could afford the poles and wires. When I was young and first visited Gilbert, I watched him walk inside his house and touch the wall, and when he did, the bulb came on. Just by touching the wall! Of course, now I know he just simply flipped a light switch. But after that day, each time I visited Gilbert and watched him touch the wall, I thought, *Why can't I touch the wall and get lights? Why am I always the one stuck in the dark, searching for a match?*

But bringing electricity to my home would take more than a simple bicycle dynamo, and my family couldn't even afford one of those. After a while I kind of stopped thinking about it altogether and focused on more important things. One of them, for instance, was graduating from primary school.

In mid-September, our teachers at Wimbe Primary finally passed out our final exams and wished us luck. For the past several months, I'd studied very hard. I stayed up late with the kerosene lamp reviewing exercise books dating back several years because the Standard Eight exam covered everything. I pored over my lessons in agriculture, remembering the proper land preparation techniques for groundnuts, the various types of farm records, and how to tell if your chickens had been stricken with Newcastle disease or fowl pox. In social and environmental science, I reviewed the roles of civil servants, politicians, and traditional authorities in the district administrative structure. Chichewa lessons were simple, so I spent much of that energy studying English, writing sentences and going over stories from my readers. One of my favorite stories was "Journey to Malwkwete," about a boy named Yembe Dodo, who's out hunting birds one morning when he's abducted by space aliens. The creatures are taller than the highest trees and wider than an elephant. Each has three eyes. Anyway, they take him aboard their spaceship and then eat his birds. I couldn't imagine such a thing.

The test lasted three days, with social studies and English on the first day, Chichewa and mathematics on the second, and primary science on the third. The three days swept by in a blur of black-and-white pages, broken pencils, and strange test-question scenarios.

I bit my nails over percentages and equilateral triangles, circumferences, and whether to apply iodine or Amprol if I found bloody layers in the henhouse. By the end I was a total mess of nerves, but felt confident nonetheless. Our grades would be posted in December, three long months away.

If I passed my exam, I'd finally be allowed to enroll in a secondary school chosen by the government. Out of six possible schools in the area, only three were boarding types. Everyone knew the government gave its best funding to the boarding schools so naturally every serious student wanted to go there. I thought it would be incredible to live on my own at a school.

But no matter where I was assigned, classes would begin in January. When they did, it would be like passing over an important threshold in becoming a man. Secondary school wasn't free, so very few Malawians even bothered attending. My older sister Annie had not only gone to secondary school in Mtunthama, but was halfway finished. I was terribly jealous of her, but now my time would come. And to me, this would usher in another important milestone: for once, I could finally ditch the schoolboy short pants of primary school and walk tall in trousers.

After I finished the exam, I waited outside for Gilbert.

"No more short pants for us, my man," I said.

"For sure, and our mornings are now free until we start school again. What shall we do?"

"Let's go hunting. It's been too long."

"Yah, for sure."

As much as I enjoyed my break from school, holidays weren't as much fun the older I became. There was always lots of work to be done on the farm, and my father needed my help. In addition to clearing the land and the ridges for the maize seeds, September was also when we prepared our tobacco for planting. Even more than maize, the tobacco seedlings require extra love and care to help them grow strong.

As I mentioned, the tobacco seedlings are first grown in nursery beds down at the *dambo* where the soil is extra fertile. Now that I was out of school, it was my job to return there each day and water the young plants from the stream, taking care to give each one the same amount of drink so it could defend itself against the sun. I would do this until December, when we uprooted the plants and transferred them to our fields.

One day in late September, after knocking off work at the nursery beds, Gilbert and I had gone to the trading center for a few games of *bawo*. Walking back to his house, I noticed something odd. About a dozen people were gathered in his yard under the grove of blue gums, talking in low voices and looking quite serious. Some men were mixed in the crowd, but mostly they were women, their heads wrapped in bright-colored *mpangos*. Each carried empty baskets. Gilbert didn't seem surprised to see them, and I asked him who they were.

"Villagers," he said. "They're running out of food in the bush. They've come to ask my father for handouts or *ganyu*. Some of them have walked for days to get here."

Ganyu was piecework, or day labor, and this was how many of the men in Malawi survived the hungry seasons when their food supply was low. Even my father had done *ganyu* in the tobacco estates, digging ridges in exchange for a few kilos of maize flour. You could usually depend on the estates in the area for *ganyu* to get you through, which is why it confused me to see people here.

I asked Gilbert, "Why can't they just work over there?"

"They've tried," he said, "but this year even the estates have nothing extra."

"So what's your father going to do?"

"He's going to feed them," he said. "He has no choice—he's their chief."

IT WAS TRUE—THE MAIZE crops in the outer villages had fared more poorly than ours during the floods and drought, and after only four months, their storages had gone dry. Soon the chatter at the trading center was that all of us were running out of food.

One day while buying salt for my mother at Mister Banda's shop, I overheard him talk about the scarcity. Each June after the harvests, Mister Banda visited several villages between here and Kasungu and bought many kilos of maize to sell during the hungry season, usually for a much higher price. But this year, the silos were empty.

"I went to Masaka and couldn't buy a thing," he said. "I even found nothing at Chimbia, and they usually have plenty. I couldn't believe what I was seeing."

I told my father what was happening at Gilbert's house, and what Mister Banda had said. He said he was already aware of this and that we shouldn't worry. Usually during the hungry season, people could go to the big Press Agriculture estate nearby and buy some maize. They grew their own food and made money each year by selling off their surplus. I told him I'd heard people in the trading center saying even Press was empty.

"Well, the government keeps a surplus," my father said. "If there's nothing at Press, the government will take maize to ADMARC and people can buy it there."

ADMARC is the Agriculture Development Marketing Corporation, a government company that sold maize on the market at discount—usually the same surplus grain from government farms. There were ADMARCS all around our district where one could go for a few kilos at a decent price.

"Don't let those people worry you, son," my father said. "Whatever the case may be, we've never gone hungry."

But one afternoon in late September, my father came home, and I overheard him talking to my mother. He'd just returned from a rally in the trading center called by the opposition Malawi Congress Party, the party of President Banda, Farmer Number One. Hundreds had attended, and the opposition officials had stood on the stage and delivered a boiling speech over a PA system. Some of President Muluzi's thugs—called the Young Democrats—had tried to stop the rally, but farmers from the villages stood guard around the stage and let the opposition speak.

These men delivered some terrible news: a few months before, President Muluzi's people had sold all our surplus grain for profit. Much of it had gone by lorry over the border to Kenya. In addition, millions of kwacha were missing, and no one in the government was taking responsibility.

"They're saying there's nothing extra," my father said. "This year will be a disaster for us all."

"We can only trust in God," my mother said.

What really happened was this: the floods and drought the previous year had given us a food deficit far greater than people realized. In addition, the international community—namely the International Monetary Fund and World Bank—had pressured the government to pay off some of its debt by selling off a portion of our grain reserve, since holding on to it was getting expensive. But some individuals in government sold all of it instead, without keeping any for emergencies. Where it all went, no one knew. Some said it went over the border to Kenya and Mozambique. Others said a large portion had been taken to ADMARC like usual, but the corrupt officials there hoarded it for too long, and it spoiled. Much of the good maize was sold to prominent traders with government connections—men who'd foreseen the food shortage and wanted to take advantage of this dire situation. They would wait until no man or woman had any food at all, and then increase the price by a hundred percent.

My father was right. We were headed for a disaster, but even he didn't know how bad it would get.

JUST AS EXPECTED, THE price of maize started going up the first week of October from its normal seasonal price of one hundred fifty kwacha a pail to three hundred. And when this happened, people began searching for other food.

One afternoon before supper, my stomach was already growling, so I went to see if my neighbor Mister Mwale had any ripe mangoes in his trees. When I arrived, Mwale and his family were sitting down to eat.

"*Eh, ndima lima,*" I said. "I've found food. Good timing, *eh?*"

Just as I said that, I noticed they were eating stewed mangoes with their pumpkin leaves. The mangoes were still unripe, green in color, and probably very sour. They were not ready to be eaten.

"You say *ndima lima,*" Mwale said, laughing. "But as you can see, we're eating the mangoes as *nsima.* You've found no food here."

Later, I saw a line of men I didn't recognize digging ridges in the Mwales' fields. They were from other villages, the same people who were gathering at Gilbert's house. When they left a few hours later, each of them carried a handful of those green, unripe mangoes as payment.

Walking through the trading center a few days later, I noticed something else I'd never seen. Several market women had spread out plastic tarps on the road and were now selling *gaga*. *Gaga* are the clear-colored husks, or chaff, that are removed from the outside of each maize kernel. Normally these husks are separated at the mill, then thrown away or sold as animal food. For me they were the perfect bait for my bird traps. *Gaga* are also used to make the *kachaso* liquor that Grandpa loved so much, and many women even burn it as firewood. I'd heard of very poor people eating *gaga* when times were tough, but it has so few nutrients that it isn't considered proper food. We feed *gaga* to our chickens, and to buy it, we have to scoop it off the floor of the mill ourselves.

But now with maize selling for three hundred kwacha a pail, I saw giant sacks of *gaga* being sold for one hundred—ten times what it had cost a month earlier. People were crowded around raising their metal pails, practically shoving one another to get it.

"Move away, I was first here!"

"We're all hungry, sister, there are no firsts in that!"

Down the road, it was the same story. When I returned an hour later, all the *gaga* was gone. Right then a kind of shock went through me, like someone shaking me awake in the middle of the night, and I began running home.

For months, my mother had been cooking our meals as if things were normal. My sisters and I got our bowl of maize porridge each morning before doing our chores. Lunch and supper was *nsima* with mustards or beans, and of course, at age thirteen, I had the appetite of a fat politician and always stacked my plate with as much as possible. Sure, I'd been aware of the drought and poor harvest and news from the opposition, but it was as if the troubles were happening to someone else.

"A little more," I'd say at supper. "That's right, keep it coming."

But after seeing people fighting for *gaga,* it was like my eyes had been opened wide and a great fear had made its way in. As I ran down the trail toward my house, I felt it grow inside me like a fist around my stomach. Once I stopped at the storage room door, it tightened its grip: out of the five bags we'd filled with grain, only two were still there. In my mind, they were already gone.

Staring at the sacks, I tried to imagine how much flour we'd get before all the grain was gone: two bags equaled six pails. One pail equaled twelve meals for my family, meaning six pails equaled seventy-two meals for twenty-four days. I then counted how many days before the next harvest: more than two hundred and ten, and at least one hundred twenty before the green *dowe* cobs would be mature enough to eat without making us sick.

Two hundred and ten days until food—and we hadn't even planted one seed. When we did, there was no guarantee it would even rain or that we'd get fertilizer. As of now, we were going to run out of food in less than a month, and I had no idea how we'd survive after that. The next time my mother returned from the mill, our flour was coarse and filled with *gaga*. Everyone began milling their grain this way, just to get a little extra.

A few days later, I saw my father rounding up our goats to sell in the market. Like many people in Malawi, our livestock was our only wealth and stature on this earth, and now we were selling it for a few pails of maize. The men who ran the *kanyenya* barbecue stands were very powerful now, and that meant they could reduce the prices however low they wanted. One of the goats was Mankhalala, a small male with long horns who'd been a favorite of mine. He'd let me grab his horns and wrestle, and sometimes even gave Khamba a good chase just to humor him.

"Papa, why are you selling our goats? I like these goats."

"A week ago the price was five hundred, now it's four hundred. I'm sorry, but we can't wait for it to go any lower."

Mankhalala and the others were tied by their front legs with a long rope. When my father started down the trail, they stumbled and began to cry. They knew their future. Mankhalala looked back, as if telling me to help him. Even Khamba whined and barked a few times, pleading their case. But I had to let them down. What could I do? My family had to eat.

In early November, I started waking up as usual at 4:00 a.m. to go make ridges in the fields. One morning, as I waited in the yard for my mother to prepare my porridge, my father stepped out into the darkness.

"No *phala* today," he said.

"Huh?"

"It's time to start cutting back. We need to stretch out what we have."

We had less than two bags of grain in the storage now, so I knew there'd be no porridge tomorrow, or the day after that. Breakfast was first to go, and I wondered what was next. Instead of complaining or asking pointless questions, I took my hoe and headed to the fields to meet Geoffrey. When I arrived, I told him about skipping breakfast.

"Can you believe it?" I said.

"You're just starting that today?" he asked. "It's been two weeks for us. I'm getting used to it."

At 4:00 a.m. the weather was cool, and I could dig my ridges with great energy. My stomach must have also been fooled from last night's *nsima,* because it had yet to wake up and grumble. But by 7:00 a.m. it was clawing and screaming to be filled, and the blazing sun was sucking all my strength. I took off my shirt and wrapped it around my head, but the extra weight of it made me more tired. The only thing keeping me from falling over was my father stomping past.

"Make those ridges better!"

"I'm hungry, out of strength."

"Think about next year, son. Try your best."

I looked down and saw my ridges were small and uneven, as if they'd been dug by a slithering snake. Across the field, my cousin swung his hoe, covered with sweat and breathing heavily.

"Mister Geoffrey," I said. "You dig my ridges today, and I'll dig yours tomorrow. Can we make this deal?"

He didn't look up. "I'll think about it," he said, gasping. "But it sounds like the same deal as yesterday."

I was trying to joke and raise his spirits, because lately I was feeling bad for him. Ever since his father died Geoffrey hadn't been the same. Sometimes Geoffrey forgot things, or he would drift off into space when I was telling him something—something of great importance, no doubt. Other times he just stayed in his room for a couple of days and didn't talk to anyone. He hadn't been feeling well, and on a trip to the clinic recently, he'd been diagnosed with anemia. I later discovered that *phala* wasn't the only thing they weren't getting at home. Food was running low all around.

"I'm joking," I shouted. "Seriously, you don't look good, man. Maybe you should take a break. Don't work so hard, and get some rest."

"I have no choice," he said, swinging his hoe. "You know my deal."

Even worse, with the recent troubles, I was pretty sure Geoffrey wouldn't be returning to school in the upcoming term, now just over a month away. His mother already struggled to pay his student fees, but now she needed him and his brother Jeremiah to work and provide food. I didn't want Geoffrey to know that I knew this, so I kept our usual banter going.

"Soon your man Kamkwamba will be in secondary school where he belongs," I said, "wearing trousers and walking tall."

"He'll find us there," Geoffrey said, smiling. "The older boys have big plans for Kamkwamba."

"What if he goes far away, to a fine boarding school such as Kasungu or Chayamba?"

"We'll find him. We have our ways."

"You can't touch him!"

"Oh, you wait and see."

Geoffrey wasn't the only one who was changing. Ever since the bad harvest, Khamba had gotten a bit slower. I hadn't realized it, but he was already an old dog when he'd first arrived with Socrates, having lived out his better years on the tobacco estate where life was good. Life in the village was much tougher, and despite the food I was feeding him each night after supper, I'm sure it wasn't enough.

As Khamba got slower, the mice began to outwit him in the fields, and the younger, faster dogs snatched up the better scraps from the rubbish piles. His thin frame got a little thinner, and I noticed he was sleeping more. He no longer chased the chickens, choosing instead to doze in the shade behind my room.

One night when I tossed up a ball of *nsima* for him to catch, he misjudged the fall and the food landed on his head.

"What's the problem, old man?" I teased. He leaned over and sucked up the food in one second flat. Some things didn't change.

EVERY MORNING AS GEOFFREY and I returned from the fields, we passed more strangers wandering the roads looking for *ganyu*. They came from Ntchisi, Mtunthama, and from the villages deep in the hills. Many carried their hoes over their shoulders with *mpango* bundles that held a cooking pot and some extra clothes.

In normal times, my mother would take our grain to the maize mill and grind it herself. She kept baskets upon baskets of flour at home so we always had what we needed. But as maize ran out all across the region, people started buying flour in one kilo or half-kilo measurements. The small portions came in small blue plastic bags and were known as *walkman,* named after the portable cassette players, because they contained only enough food for one person. Walkman was for city people, not farmers and their families.

By now, mangoes were out of season and gone, so many men also worked in exchange for a few leaves of cassava, or manioc. This plant is a root vegetable whose leaves can be stewed like spinach or rape, and whose tubers can be dried and pounded to flour. Cassava is quite popular in other parts of Africa, especially Congo. But in Malawi, cassava was like *gaga*— rejected in better times when there was plenty of *nsima*.

"*Ganyu* for walkman? *Ganyu* for cassava?" the travelers now cried.

Many people were headed to our district to find work on the estates. But little did they know, most of the estate employees were hanging around in the market, out of jobs and hoping for miracles.

"Excuse me," these outsiders would ask. "Where can I find Estate 24?"

"Don't bother, friend," the workers answered. "There's nothing. And I should know—I work there."

As the hungry *waganyu* made their way through the market, they passed by the homes of the wealthy traders, whose windows released the smells of stewed chicken and *nsima* at lunchtime, as if the belly of the world wasn't howling to be filled. Because so many farmers were selling off their animals, chicken was practically free for people with money.

While the men looked for work, their wives and children gathered at Gilbert's house hoping for a handout. About forty arrived each morning, and hundreds of others had passed through since the troubles had begun in August. Most stayed long enough to receive a walkman from Gilbert's mother, while some arrived too weak to continue. They spread their blankets under the trees and cooked their *nsima,* leaving only after they had regained some of their strength. Others collapsed on the road and had to be cared for. The blue gum grove had become a patchwork of Gilbert's family's bedding.

AROUND THIS TIME, PRESIDENT Muluzi was busy traveling the country in his usual fashion, giving out small handouts of money and showing that he was a Big Man. Massive rallies were held for loyal officials, complete with dancers, military parades, and lots of food. Everywhere the president went, he tossed the poor just enough flour or kwacha to make sure they remembered him come election day.

I'd seen him at Wimbe Primary School in 1999 during one of his whistle-stop tours. There'd been women's choirs and Gule Wamkulu, with each dancer receiving fifty kwacha notes. The local politicians had also lined up to get their handouts. This was the first time I'd ever seen the president in person. He was bald and fat, and when he stood up and walked to the podium, his short legs seemed mismatched for his round body.

He'd said something like, "I'm ashamed to see this school broken in such fashion. We should tear the whole place down and start from scratch, build it again strong and proud! Teachers' houses also need to be ship-shape, and students need new desks and books!"

Of course, the crowds cheered and applauded at this. But instead of buying us new desks, he sent men into our blue gum grove to chop down our trees to build them. Even then, there weren't enough. The teachers never got new homes, and the only thing he did for the school building was give it a fresh coat of paint and a new iron sheet for the roof.

This was the same president who'd promised that every person in Malawi would get new shoes if he won the election. Well, you can imagine. After the people voted and Muluzi claimed his victory, everyone started asking, "Where are our shoes?" The president went on the radio and said something like, "Ladies and gentlemen, do I look crazy? How can I know the shoe size of every person in Malawi? *I never promised shoes.*" Our president was a funny guy.

Despite a growing anger about the missing maize surplus, the government said nothing on the radio. And despite the looming hunger, it offered no solutions. So when President Muluzi announced he was stopping in Kasungu to appoint a local chief, all the subchiefs from the district pleaded with Gilbert's father to stand up and speak on their behalf. Like the president, Gilbert's father was a Muslim, and that could work to his advantage.

"You're a good speaker and one of his people," they told him. "You must convince him to save us."

The day of the rally, several thousand people stood in the sun, hoping to hear what the president had to say about the crisis. But instead of answers, they got dancing and speeches that dragged on for hours, speeches about how the president was a great and powerful man, and how he was kind enough to approve new development in the area, such as building new toilets in some villages and digging a few wells.

At the time, the president was also chairman of the Southern African Development Community, a kind of social and economic alliance between fifteen countries in southern Africa. During his tenure, there had been terrible wars in other parts of the continent, places like Angola, Burundi, and Sudan. The genocide in Rwanda that killed more than eight hundred thousand Tutsi had then spilled into the Democratic Republic of Congo and started a war there. Muluzi had even hosted the leaders of Congo and Rwanda in Blantyre in an effort to broker peace. Given his good work across Africa, it was confusing why he failed to see our problems here at home.

Anyway, after an hour of dancing and singing, it was finally time for Gilbert's father to speak. He left his seat, which was just in front of the president, and stood at the podium before the people.

"Your Excellency," he said, turning to face the president. "I'd like to congratulate you not only for what you've done in Malawi, but all across the great continent of Africa. We're hearing about all the things you're doing in Congo and how you've had success. We're very proud of our president. But please understand, we're also at war here in Malawi, and that war is against hunger."

He then asked the president to stop funding wells and toilets and use the money to buy grain. (Because really, how can you use the toilet if you never eat?)

The crowd erupted in applause that lasted so long that the next speaker had to stand there and wait until it was silent. When the crowd simmered down, this next speaker only heaped praise on the president, and the audience began to boo and hiss.

"Sit down, you have nothing to say!" they shouted.

"Chief Wimbe already spoke for us!"

"Foolish politician! There's no politics in *nsima*!"

Shortly after, when the president got up to talk, several well-dressed officials approached Gilbert's father and asked to speak with him. Knowing the president's habit of giving handouts, the chief became excited, thinking, *They're giving us money. My speech must've worked.*

About six men led the chief behind a building near the stage, and once there, they confronted him.

"In what capacity were you speaking such nonsense?" one asked, looking very angry.

Before Chief Wimbe could answer, they knocked him to the ground and began beating him with clubs and batons. After several minutes, the thugs left the chief bleeding in the dirt and slipped away into the crowd. When a friend discovered Chief Wimbe a while later, he refused to go to the hospital, fearing the thugs would kill him in his bed. That afternoon when Gilbert came home, he discovered his father lying on the sofa, unable to move. By that evening, large black and purple bruises were visible all over his chest, stomach, and arms.

For the next several weeks, the chief remained on the sofa and in bed, trying to recover from his wounds. He then began sneaking to clinics across the district, keeping his whereabouts secret. After many tests and treatments, he never told anyone the results. Fearing Muluzi's people would discover his actions, Chief Wimbe suffered in silence.

For me, this turn of events was frightening. Our chief was like our father, the man who protected our small area and represented us to the rest of the country. When we heard he'd been beaten, it was as if we'd all been violated, our safety no longer guaranteed. If the government treated our dear leader in such a way, with the hunger bearing down, I wondered if we people would fare much better.

CHAPTER SIX

ECEMBER ARRIVED WITH HEAVY clouds, black as oil, that gathered for days over the village before finally releasing their rain. All across the district, farmers did their best to plant seed for the next harvest, yet many were so busy looking for food that their fields went unsowed. Our family was fortunate that we were able to plant a small crop of maize. We also managed to plant a half-acre of tobacco, which would be a lifesaver in months to come.

Each day while I weeded the fields, I'd see the *waganyu* moving slowly down the road searching for piecework, their clothes soaked with rain and covered in mud. Working for food was becoming harder now that the price of maize at the market was increasing every day. Three hours of work yesterday became six hours today, all for the same bag of flour.

The *waganyu* and others continued to gather at the home of Chief Wimbe, whom they knew would have extra food in storage, for the chief's maize fields were huge and planted with plenty of fertilizer. It became Gilbert's job to greet the hungry people and help his mother pass out *phala* at the back door. After he'd served one person, a new one would appear.

"*Odi, odi,*" they'd say. "Anyone here?"

"Another one," said Gilbert, "this one worse than before."

Each day as they left Gilbert's and continued down the road toward our house, I imagined my father and me walking beside them, heads toward the ground, scouring the soil. I feared it wouldn't be long. Back at home, our food was nearly gone.

Just the previous day, my mother had milled our last pail of grain, and I knew that meant only twelve more meals. When she'd gone, I opened the storage door and peered inside. The empty sacks sat in the corner like a pile of dirty clothes for the wash. I tried to imagine what the room had looked like when it was full, when our lives had been normal and fear didn't live inside us. But I couldn't summon the energy to remember.

That night, my father called us together in the living room.

"Given our situation," he said, "I've decided it's better if we go down to one meal per day. It's the only way we'll make it."

My sisters and I understood, but still argued over the fine points.

"If we have one meal a day, when should it be?" asked Annie.

"Breakfast," said Aisha.

"I like lunch!" shouted Doris.

"No," my father said. "It will be supper. It's easier to keep your mind off hunger during the day. But no person should try to sleep with an empty stomach. We'll eat at night."

So starting the next evening, we ate our single meal of the day. Again my father gathered us all in the living room, making it the first time we'd ever eaten together as a family.

In our Chewa culture—at least in the village—the daughter *never* eats with the father, and the son *never* eats with the mother. It's not considered polite (what if you pass gas in front of your mother?). As early as I remember, I'd always eaten with my father and uncles, while my mother fed my sisters in a separate room.

In our culture, family life follows many rules that have been passed down by our ancestors. Family relations aren't like they are in America, where daughters hug their fathers and sons hug their mothers. If people saw such a thing in the village, they'd wonder, "Where are their morals?" And children respect their elders of the opposite sex, no matter what. For instance, if I called my younger sister over and held out a hundred-kwacha note, saying, "Run quickly to the shop and get some bread," she'd bow to one knee before taking it. It's like that.

So that night in the living room, my father and I sat on the floor with my sisters. A Nido can filled with kerosene flickered at the end of the wooden table, sending black soot spiraling through the humidity. My mother came in carrying a basin of warm water and a pitcher for hand washing, which is done before and after every meal. My sister Doris went around to each person, holding the basin under their hands while pouring with the other. When the hand washing was finished, my mother fetched a large bowl, then lifted the lid.

"Try to make it last," she said, then joined us on the floor.

Instead of the mountain of *nsima* cakes I was so accustomed to, the bowl contained only one gray blob. It didn't even look like food. Another bowl nearby contained some mustard greens. Soon the smell found our nostrils and we passed the blob from one person to the next, picking it apart like a yardful of hens. We didn't even bother using plates. The way I calculated it, each person got seven mouthfuls of food. We finished the *nsima* in minutes and ate mostly in silence, happy that at least we were chewing something.

But there were no signs of happiness on my parents' faces. In fact, I'd never seen them so terrified in all my life. You see, the week before we ran out of food, on November 22, my mother had given birth to another baby girl.

IF OUR CULTURE DEMANDED that children respect their elders, it also forbade us from asking questions, especially about things concerning the body. I'd noticed my mother getting fatter for months, but didn't dare say a thing. When a woman gets pregnant in the village, it becomes taboo, an open secret never to be discussed. Only the woman's husband and mother can ask her about her growing belly. And if a child was overheard even mentioning this aloud, he or she would surely be beaten. Talking about a pregnant woman was not only none of your business, but people believed it also left her open to a kind of bewitching. Many pregnant women simply stayed indoors until the baby was born. When youngsters asked where their new brother or sister came from, the parents would say, "The clinic, where all children are bought."

So when my parents came home with a new baby girl, my young sisters jumped with excitement, knowing they too had been purchased at the clinic. But my parents seemed too worried to even entertain their questions. For days, my new sister didn't even have a name.

In the villages where health care is poor, many children die early of malnutrition, malaria, or diarrhea. In hungry times, the situation is always worse. Because of this, names often reflect the circumstances or the parents' greatest fears. It's quite sad, but all across Malawi, you run into men and women named such things as Simkhalitsa (I'm Dying Anyway), Malazani (Finish Me Off), Maliro (Funeral), Manda (Tombstone), or Phelantuni (Kill Me Quick)—all of whom had fortunately outwitted their unfortunate names. Many change their names once they're older, like my father's older brother. My grandparents named him Mdzimange, which means "Suicide." He later changed it to Musaiwale, meaning "Don't Forget."

Despite the stress my parents were facing, my sister was born healthy—six pounds, two ounces. Whether it was due to her good health, or a kind of blind faith as we entered a famine, my parents named her Tiyamike, which means "Thank God."

WITH ONLY HALF A pail of flour remaining, I knew it wouldn't be long before we'd join the *waganyu* and begin roaming the land. We needed some kind of miracle, or at least a good idea. The next morning, my father announced a brilliant plan—a gamble, a roll of the bones even riskier than magic.

"We're selling all our food," he said.

That same morning, my mother took our last bit of flour, mixed it with soy and a bit of sugar, and began baking *zigumu* cakes to sell in the market. The plan was to jump-start a small business, take advantage of the scarcity, and live on the profits—if there were any.

All day the smell of sweet cakes filled our compound, invading every room and drifting out into the fields. Several people stopped in the road, maybe hoping for something, and just took in the aroma. Even the birds became brave and gathered in the yard to sing a woeful tune. The smell seemed to enter my body like a spirit, slithering into my empty belly and stretching its legs, using its elbows to get comfortable. It was torture. Normally, my mother let me take the bowl and scrape the remaining batter with my fingers. Leftovers were so cherished that we kids had even given them a name: VP, after "*vapasi* pot," meaning the bottom of the pot. We'd appear in the courtyard as my mother was preparing to wash the pot or give the scraps to the chickens, saying, "Mama, VP?" But this time was different. My mother had scraped away every last drop to use, as if wiping it clean with a sponge. No VP, only the bare pot.

That night my father built a crude stall from a broken table and an iron sheet, and he and my mother positioned it in front of Iponga Barber Shop. The next morning, my mother opened for business, selling her cakes for three kwacha each. The cakes were heavy and filling and cheaper than the buns and walkman that were also for sale. If a person had just some small change, but not enough to afford a bag of flour, the cakes were the only option. Some days, my mother sold out in less than twenty minutes.

Ever since the country had run out of maize, businessmen in the trading center had been crossing the border into Tanzania and buying it by the ton, then marking it up to sell. One of these traders, Mister Mangochi, was an old friend of my father's. With the money my mother earned from selling cakes, my father cut a deal with Mangochi and bought one pail of maize. My mother took it to the mill, saved half the flour for us, and used the rest for more cakes. We did this every day, taking enough to eat and selling the rest. It was enough to provide our one blob of *nsima* each night, along with some pumpkin leaves. It was practically nothing, yet knowing it would be there somehow made the hunger less painful.

"As long as we can stay in business," my father said, "we'll make it through. Our profit is that we live."

On a Sunday morning not long after, when my mother was home preparing her things to sell, she noticed something odd. Two young men on bicycles were standing in our yard talking with my older sister Annie. My mother had never seen them before, and since my sisters weren't allowed to speak to boys without permission, she went over to investigate.

"Mama, these are two teachers from the private school in Mtunthama," Annie said. "They've come to visit one of their friends down the road."

Annie asked if she could escort them. Annie attended the secondary school just across from the private school, so my mother suspected they all knew one another. She agreed, thinking nothing of it, and went off to the trading center. At the time, my father was visiting a friend at a nearby estate, and since it was Sunday, the rest of us had gone to the market. Only my sister Doris, who was nine years old, stayed behind to take care of the house.

When my mother returned home that afternoon, she discovered that nothing had been done to prepare supper.

"Why isn't there a fire made?" she asked Doris. "Where's your sister?"

"She went with the men."

"She's still not back yet?"

Doris shrugged her shoulders.

That night when my father returned home, he asked where Annie was. My mother said she didn't know. She didn't want to tell him about the boys, hoping Annie would just come home before it got any later. But after supper, when Annie still hadn't returned, my father asked again. This time he seemed angry.

"Where is my daughter?"

"I don't know."

"Of course you know. You're her mother. Now tell me—"

"*Please,* I don't know."

My mother became so worried she took her torch and went looking on the roads. She asked the neighbors, the people coming from the market, but no one had seen Annie. After some hours, she came home in tears and confronted Doris again.

"What did you see? Where did they go?"

Frightened, Doris then told my mother the truth. Just before Annie left with the two young men, she had packed a bag. Annie told Doris not to tell.

My mother rushed into Annie's room and saw that all her clothing was gone, along with her schoolbag. The only thing left behind was her school uniform and books. Turning, my mother noticed something sticking out from Tiyamike's diaper bag, which she kept in Annie's room. It was a note, written in Chichewa: "I GOT MARRIED TO THE TEACHER. I'M SAFE, DON'T WORRY."

My father came into the room, and my mother read the note aloud. When he heard this, he flew into a rage.

"*Who is this man?*"

"I don't know."

"*Tell me right now who he is. I'll go and find him!*"

"I don't know who he is!"

I could hear my father stomping up and down the yard, his nostrils flaring like an angry ox. I didn't dare leave my room, for fear he'd whip me just for being there.

"*You're lying! You're hiding her from me! Where is my daughter?! Go and get my daughter back. You know where she is!*"

"I told you I don't know!" my mother said.

Annie had just passed her Form Two exam and was preparing to enter her junior year in high school. My parents had been so proud, bragging to all the relatives and traders in the market. It was common for my father to sit my sisters down and tell them things like, "I saw a girl working in the bank in town, and she was a girl just like you." My parents had never completed primary school. They couldn't speak English or even read that well. My parents only knew the language of numbers, buying and selling, but they wanted more for their kids. That's why my father had scraped the money together and kept Annie in school, despite the famine and other troubles. Now she was gone without even saying good-bye.

"I've lost my daughter and all my money!" my father cried. "My daughter is a stupid girl. Wait until I find her!"

But he knew it was too late. He knew Annie was probably staying with her lover that night. Even if the arrangement failed and they didn't get married, my sister had dirtied herself and shamed our family. She could never live at home again, and this crushed my father's heart.

The man Annie married was a teacher named Mike. They'd met in Mtunthama months before and had fallen in love. Mike had actually visited the day before without anyone seeing and told my sister, "I'm coming here to pick you up. I don't want you living in this village anymore." The friend had come along as a ruse and taken the bag ahead. That way, no one would suspect anything if they saw Mike and my sister together. The two hid for the night at the friend's house at Estate 34, then went to Mike's home in Ntchisi the following morning.

In a normal courtship, if a girl likes a guy enough, she'll ask him to come meet her family. He'll then visit for several weekends in a row. If all goes well and the guy senses he's safe with the family, he'll propose marriage. The girl will say, "Okay, let me go talk with my uncle." The girl will then speak to her mother about the matter. The mother will talk to the father, who will speak to his wife's brother. The uncle then approaches the groom's uncle and the two men convene. The bride's uncle then names the price of the dowry—usually money, anywhere upwards of one hundred thousand kwacha, or livestock, such as a cow. The bride's uncle will also be given a "mouth fee," for speaking on behalf of his family. All of this happens before the wedding is even planned.

In addition to the dowry, the groom's family also pays for the ceremony and reception: all the food, drinks, transport—spending big if he's any kind of *bwana*. For a man, weddings are expensive, which explains why there are so many young, single men in Malawi.

My sister Annie's fiancé did none of this. Three weeks after she disappeared, my father received a letter from Mike's parents informing my parents of the marriage, and how the couple planned to stay in Ntchisi. It then gave instructions on where to collect the dowry, just a few hundred kwacha, of which only half arrived. It would be more than a year before we heard from my sister again.

MY FATHER'S SPIRIT SEEMED to disappear in the days after my sister left. He was no longer his upbeat, optimistic self, and his laughter was noticeably missing from the compound, as if the wind had come along and stolen one of our houses. He seemed to grow increasingly quiet and sullen, which did little to help our confidence in the face of looming hunger. We were all upset about Annie, but what we didn't admit was that her absence now meant a little more food. With her gone, each person got an extra mouthful at supper.

About a week later, my mother was coming home from selling cakes in the trading center when a thirty-ton lorry passed her on the road. Its load was covered with tarps. A few traders nearby said it was full of maize and headed for the ADMARC in Chamama. When my mother got home, she told us the news and called me over.

"You will go to Chamama tomorrow. Leave as early as possible."

Chamama was fifteen kilometers away, so of course I grumbled.

"Are you positive it was maize and not fertilizer, because if—"

"Are you listening to me, boy? You go tomorrow."

If my mother was right, this was great news. Maize was now selling for eight hundred kwacha a pail in the trading center and was sure to increase even more in the coming months. If we could get it cheaper at the ADMARC, then the extra bag or two would help us a great deal.

The next morning I awoke at 5:00 A.M., got on my bicycle, and headed for Chamama. I took the shortcuts through the fields to get there faster, but it didn't really matter. The trails were full of people going in the same direction. All of us carried empty flour sacks.

"Chamama?" I shouted.

"*Ehhh*," they answered.

The ADMARC was located in the center of Chamama's trading center, along a gravel road amid a row of white, one-story storefronts. The building itself was white with blue trim along the windows, housed behind a wire fence and surrounded by metal sheds that normally held all the grain.

I couldn't believe what I saw when I got there. The lines stretched from the doors of ADMARC all the way down the road, as long as a soccer pitch at least. One line was for men, the other for women, and each became longer by the minute. I parked my bike against the fence and went and took my place.

At 6:15 A.M., the sky was still a bit dark. The weather was cool and pleasant and people seemed in good spirits. But once the scorching sun rose in the sky, I suddenly became aware of how devastating the hunger was for everyone. All around me people appeared weak and exhausted, as if they hadn't slept. The skin around their cheeks was shrunken, and their eyes were pinched against the light. They probably hadn't eaten much in weeks, and I suspected the ADMARC was their only salvation. If things went badly for them here, they probably wouldn't get any better. As the air became thicker and hotter, they began to wilt like potted plants in the sun.

An old man in front of me could hardly stay awake. His hands trembled as if he was cold, and his breathing was heavy. When the line advanced a step, his body couldn't follow, and he collapsed to the ground. To my horror, the crowd simply stepped over him. In the next line, babies cried and wailed from hunger, and children tugged on their mothers' dresses, begging for breakfast. If there's anything I remember most about that day in Chamama, it's the sound of crying babies.

As the morning dragged on, people began selling their places in line. These men had no money, but they had arrived around 3:00 A.M. and secured a good spot in line. As they neared the building, they turned to the person behind them and said, "Hold my place, *eh*?" then walked to the back of the line. They found the person shaking most from hunger, the man whose eyes were dim and dancing in their sockets, and told him, "I'm ten minutes from the door. You can have my spot in exchange for a little maize on your way out." People jumped at the offer. This happened all day.

That morning, my mother had dipped into her inventory and given me a cake for the journey, which had sat like a stone in my empty stomach. But after many hours in line, I became very hungry and weak myself. The heat from the other bodies was like being surrounded in fire. People began smelling bad, and my head felt like a plastic *jumbo,* caught in the wind and drifting into the sun. Even my fingers were sweating.

As we inched closer to the door, people became impatient and began pushing. They simply couldn't wait any longer. Someone pushed me so hard from behind that I had to grab hold of the man in front of me just to keep from falling down. Soon people began running from the back and squeezing in front, twisting and squirming into the line like mice under a door.

"*Eh,* stop cutting!" people screamed, the hunger straining their voices.

"We woke up with the first cock! We've been here all day!"

But people still kept coming. Everyone knew there was only one truck of grain, and it would run out eventually. The more people cut in line, the more the others panicked.

Both lines mobbed the front doors. The wave of bodies surged behind me, pinning me against the man in front. I couldn't breathe. I planted my feet to keep from falling, but when I leaned sideways for some air— no more than three inches—the man behind me tumbled forward and sparked a chain reaction. Four men collapsed on top of him, causing a great crash of knees and elbows. Once the dam had broken, the crowd rushed over them to fill the empty space.

As the mob swallowed me again, a strange thing happened. Everything went quiet. The screams and moaning of children fell away, along with my own fear. I looked out through the tangle of faces, fingers, and teeth and saw my great reward. The ADMARC building lay just ahead, only a few short meters away. A shallow rain ditch circled the building like a castle moat, and in my misery, that ditch became like the great river Jordan. There I was, standing atop the mountain, peering into the Promised Land. All I had to do was make it across the water.

Twenty minutes later, with a few more surges of the crowd, I felt my feet clear the distance. A second later, I was inside. The ADMARC office was quiet and peaceful and the air suddenly became cool and fresh. Ahead of me, I saw a hill of maize as high as my waist, more food than I'd seen in months. The sight was intoxicating.

But just as I stepped inside, I heard more commotion near the door, where I'd just been standing. Two workers now stood before the crowd and made an announcement: "Ladies and gentlemen," one said, "we're sorry, but we have only ten bags left—"

The crowd didn't wait for him to finish. The riot I'd just escaped now exploded into one giant fight. Men used their fists to claw their way forward. I saw a woman fall to the ground and vanish. Another man was dragged to the dirt and trampled. Just near the door, several women carrying infants jumped out of the line to avoid being crushed. These women had been there since sunrise, but by pulling themselves out of the fracas, they also lost their places. I watched them walk away with nothing, wondering if they'd make it through the month.

Turning away from the violence outside, I realized it was my turn in line. I had four hundred kwacha in my pocket. It was enough to buy twenty-five kilograms of maize, which was the advertised limit. But now at the seller's table, I was told I could purchase only twenty. The price remained the same.

"So how much do you want?" the man asked.

"Twenty."

He gave me a ticket and pointed down the line. At the other end, workers used metal pails to scoop the maize. The men looked muscular and healthy, nothing like the people outside. The worker who measured my maize then cheated me. He filled my pail very quickly then tossed it onto the scale, causing the needle to bounce wildly from one end to the next. But before it could settle, he whipped it off and emptied it into my sack.

"*Next!*"

"But wait! You didn't—"

"If you don't like it, you can leave your maize here! There are plenty of people behind you. *Next!*"

With little else to do, I handed him my money, grabbed my maize, and ran for the door, as if I'd just robbed the place. Despite being cheated, I felt like I was on top of the world, the jackpot winner. But my joy quickly turned to fear as I stepped outside into the mob.

A man rushed toward me, shouting, "I'll give you five hundred for that!"

Another pushed him aside. "No, boy, I'll give you six hundred!"

I pretended not to listen, concentrating only on strapping the bag to my bike and getting away. The next person could easily just beat me and steal my food. Once I reached the road, I never stopped pedaling. Later, we heard that there were riots at other ADMARC buildings across Malawi, where several children fell from their mothers' backs and were trampled.

At home, my sisters and parents welcomed me back like a hero from the war. I must've looked exhausted, and my clothes were stretched and dirty. I tossed the bag onto my father's scale and confirmed I'd been robbed. I'd only received fifteen kilos of maize, half a bag, but at least that would feed us for another week.

Soon after I returned from Chamama, people began selling their possessions.

Standing on the porch one morning during a heavy rain, I watched a line of people walking slowly past like a great army of ants. They were our neighbors and farmers from other villages. The women carried large basins atop their heads containing the items from their kitchens, all their pots and pans, water buckets, even bundles of clothing. A man carried a chicken under each arm. Goats were bound by the leg with a vine, crying and *baaing* as they were dragged through the mud. Men balanced sofas, chairs, and tables across their backs and shoulders, their heads bent and faces twisted under the weight. They stopped to rest every few meters, exhausted from hunger, before hoisting their heavy loads and continuing on.

At my feet, Khamba lay spread out on the ground, lazily flipping his tail at the flies that had sought shelter under the porch. He was growing even thinner. Our one meal a day didn't include my dog, and some days, I was forced to eat what I'd normally give him. I felt so guilty that most of the time I tried to avoid him completely. But this morning he'd found me, and together we watched the people on the road.

They all seemed to be in a great rush to get somewhere, to unload these worldly items and put something in their stomachs. They hardly seemed to notice the rain, which had become so heavy their bodies blurred into watery spirits.

I waited for the rain to taper off, then followed them down the muddy trail to the trading center. The rainy season normally doesn't deter the market women, who stand out all day under giant umbrellas and never get wet. But today their wooden stalls in the market were empty. Many of the shops along the main road were also shuttered, but this was no ghost town. As a light shower fell, the market women and most everyone else now crammed the main road, looking for food. Most normal, everyday trading was now being replaced with the business of survival.

Usually people who wanted to sell their possessions would spread out tarpaulins to display what they had. But now they walked from person to person, saying, "*Ndiri ndi malonda*. I've got something to sell. How about this radio? It's yours for a giveaway price."

Those carrying furniture weren't able to squeeze under the awnings for shelter, and the rain soaked the tables and sofas on their backs. But still they continued on.

"Don't worry about the water," said a man with a nervous grin. "This is hardwood, it won't ruin. You'll have this chair into your old years. How much do you have? I'll take anything. My children need to eat."

A few of the businessmen like Mister Mangochi bought things they later gave back. But most people had no money. They simply shrugged and shook their heads.

Crowds now gathered around the few traders who were selling imported maize. The prices were so high that the traders were practically regarded as criminals.

"You people are thieves!" people shouted.

"Who's setting these prices?"

"You're killing our children!"

Soon people began selling the iron sheets off their homes for a cup of flour, and their thatched roofs for even less.

"What good is a roof when I'm dead?" one man asked.

A man in the trading center was caught trying to sell his two young daughters. The buyer had informed the police. People were becoming desperate.

By now the December rains had caused the grass to grow very high by the roadsides. And since people were becoming weak or were out looking for food, the grass went untrimmed. The tall grass made a perfect hiding place for thieves, and soon women were being attacked and robbed as they left the maize mill. One morning as I walked home along the mill road, I saw a young mother standing alone, sobbing. The crime had just occurred.

"My children are waiting," she cried. "What am I going to do?"

A few other women appeared and tried to console her.

"We know your kids are crying at home," they said. "Just send your husband next time."

"Next time?" the young mother replied. "There may not be a next time."

The mountain was getting higher for everyone.

Inside the maize mill, the owners no longer had any use for a broom. The hungry people kept the floors cleaner than a wet mop. At the beginning of the month, the mill was packed full of those waiting for fallen scraps. The crowd would part long enough to allow women to pass with their pails of grain. As the machine rumbled and spit a white cloud of flour into the pails, the multitude of old people, women, and children watched intently with eyes dancing like butterflies. Once the pail was pulled away, they threw themselves on hands and knees and scooped the floor clean. Afterward, old women would rattle their walking sticks up inside the grinder as if ringing a bell, collecting the loose flour that drifted to the floor.

This activity stopped by mid-December because so few people had any grain to mill anyway. The building remained empty, all except the operator and a few children whose parents had either died or had simply abandoned them to hunger.

NORMALLY, CHRISTMAS WAS MY favorite day of the year—but of course, this year was far from normal. In better times, we'd celebrate Christmas Eve by attending the nativity play at the Catholic church down the road, watching Joseph and Mary and Baby Jesus try to escape from Herod's soldiers and their wooden swords and AK-47s (it wasn't the most accurate version, but it was funny).

After church, we'd feast on delicious flying ants that arrived with each rainy season. The insects would swarm the lights of the trading center at night, then ceremoniously fall to the ground and shed their white wings. But since we kids weren't allowed out after dark, my sisters would build a large grass fire behind the house, then catch the ants in basins of water as they fell. The drowned ants were then roasted on a flat pan and doused with salt. Roasted ants tasted like chewy dried onions and stuck nicely to a ball of *nsima*. When eaten along with beans and pumpkin leaves, they were truly heavenly.

On Christmas morning, breakfast wouldn't consist of *phala*, but sliced brown bread, soft and moist. If we had extra money for other luxuries, such as Blue Band margarine, sugar, and powdered milk, I'd press several pieces of bread together into a buttery sandwich, then wash it down with a mug of steaming Chombe tea. Fresh bread and Blue Band, mixed with sweet and milky tea, is one of the best combinations you can ever put into your mouth.

Malawians' desire for meat is especially strong at Christmas. If you go all year without eating meat, hopefully on Christmas Day you can manage a little something. So in the early afternoon on Christmas Day, my father would kill one of our chickens. Christmas chicken isn't served with *nsima*—not today—but with rice. Ask any Malawian to talk about Christmas, and he'll always mention the rice.

But Christmas of 2001 arrived more like a punishment. First of all, the week before the holidays, most of our chickens got Newcastle disease, which paralyzed them, made them unable to eat, and finally killed them. Only one chicken managed to survive, and I'm telling you she was very lonely. The Catholic church canceled the nativity service, and the Presbyterians didn't even bother making an announcement. But no one showed up. And with everyone so hungry and getting weaker, my sisters couldn't even muster the energy to catch flying ants.

On Christmas morning, there was no fresh bread or Blue Band waiting on the table, no tea with milk and sugar, no chicken for supper, and certainly not any rice. There was no breakfast at all. I woke up, washed my mouth, and heard the sounds of "Silent Night" coming from the radio in my sisters' room. When the song ended, the young DJ came on, sounding very energetic.

"*Ehhh,* here's wishing all of you a *very merry* Christmas," he said.

"Easy to say when you're in Blantyre, working for the government," I said, then grabbed my hoe and headed for the fields, anything to keep from hearing more about Christmas.

Around noon, my mother did manage to serve a Christmas lunch, but it was just our usual blob of *nsima* and a spoonful of pumpkin leaves. She'd probably worked very hard to save that much flour to give us an extra meal, but it was hard to feel festive. I finished my last bite and was still hungry.

I then went to see Geoffrey, which turned out to be even more depressing. I walked into his room and found him sitting on the edge of his bed, looking exhausted. Ever since Geoffrey's mom had milled their last pail of maize back in November, he had become one of the *waganyu* who wandered the land looking for piecework. He'd been fortunate to find jobs pulling weeds and bunking ridges in a trader's field near the *dambo*. I could see that his anemia was still bothering him, and he was losing weight.

"*Eh*, man," I said, "I haven't seen you in days. Your field is full of weeds. They're taking over."

"I'm too busy with *ganyu*," he said. "At first, I went out looking for food for the month, then the week. Now it's all about tomorrow. How can I tend my fields when I can't even eat tomorrow?"

I wanted to help him, but what could I do? I hadn't seen Gilbert in a few days, so I walked down to his house. It was creepy along the trail. Usually on Christmas you'd hear the sounds of music or singing and see families dressed in their best clothes laughing as they made their way to the trading center. But the people who walked past today did so slowly, heads down, not even saying hello.

About fifty people were scattered under the blue gums outside Gilbert's house when I arrived. The smoke from their cooking fires covered the house in a gray haze. Gilbert was standing in his doorway.

"Merry Christmas, *eh*?" I said.

"Not here," he answered, shaking his head.

Surely Chief Wimbe will have some delicious chicken and rice, I thought. But Gilbert said no, the nonstop procession of souls to his door had taken most of their supply.

"It's only *nsima* and beans here, brother," he said, looking disappointed.

"How's your father?" I asked.

"He greets people when he's feeling well," he said. "But mostly he sleeps and listens to the radio with his cat."

My nose then caught the scent of something terrible on the breeze. It made my lips curl.

"What is that?" I said.

"Oh, the people aren't even bothering with the latrine anymore. Now they're just defecating in our trees. Be careful where you walk."

"Yah, for sure."

With Gilbert busy with the hungry people, and Geoffrey not feeling well and doing *ganyu,* I went to the clubhouse to see what my cousin Charity was doing. I knocked on the clubhouse door, and Charity let me inside.

He had a small fire going in the pot. His roommate Mizeck was nowhere around.

"It's Christmas," he said, "and I haven't eaten a thing."

"Yah," I said. "I'm hungry, too."

The two of us began thinking of ways to get food. The mangoes were all gone, so we couldn't steal those. The traders in the market wouldn't dare part with any flour. And we weren't hungry enough to crawl outside their shops and dig for stray kernels in the dirt—not yet anyway.

"We need meat," Charity said. "I can't go to sleep tonight without my Christmas meat."

There was a guy named James who ran a *mang'ina* stand near the *kanyenya* pit, where meat is fried. *Mang'ina* is the meat from the head and hooves of a cow or goat, basically headcheese. The butcher splits the head into three or four parts, then tosses them into a pot of boiling water, along with the legs and the hooves. You could come by and eat the tender meat from the leg, or have a bit of boiled brain and tongue. The meat from a cow's cheek was also delicious.

Charity and I started dreaming aloud.

"Perhaps James will be generous on Christmas and let us have some."

"Don't be stupid," said Charity. "He'll never do that."

He paused for a moment, then said, "But he does throw away the skins."

"Can you eat that?" I asked, twisting my face.

"I'm thinking why not? What's the difference? It's all meat, right?"

"Yah, I guess you're right," I said.

The hunger had affected our thinking.

As we walked to James's stand, we saw that the *kanyenya* boys were doing brisk business as usual. Despite the hunger, the wealthy traders were crowded around their stand chewing on grilled meat, not even swallowing before shoving in a handful of fried chips. A group of villagers were crowded around watching them eat, studying the motions of the traders' hands as they dipped the greasy bits of meat into the salt before popping them into their mouths. Watching them chew, I could feel the salty burn on my own tongue.

James's stand was just a bit farther down the road. He was there, as usual, standing above a giant pot boiling on the fire. Getting closer, I could see a delicious goat head and some leg pieces swimming around inside. I wanted to leave immediately.

"*Eh,* James," said Charity. "William and I are making a Christmas drum for the children in the village. Can you part with one of your skins?"

James looked up from his pot. "That's a good idea," he said, then turned around and nodded toward a mound of something on the ground. It was heaped atop a black plastic *jumbo* and crawling with flies.

"I have this goat skin," he said. "I was going to throw it away, but you can take it."

Charity quickly threw the skin inside the bag and handed it to me. It was still warm.

"*Zikomo kwa mbiri,*" Charity said. "The kids will appreciate you."

"Sure, sure."

With our warm goat skin in hand, we hurried back down the hill to *mphala.*

"How are we going to prepare this?" I asked.

"It's easy," he said. "We'll just do it like a pig."

Back at the clubhouse, I lit a clump of blue gum bark and got the fire started again, then added a few small sticks, since we'd run out of maize piths long before. When the fire was strong and hot, Charity and I held the corners of the skin and stretched it flat over the flames. Soon the heat was singeing the hair, curling it black. Normally this would make a terrible stench, but now in my hunger, all I smelled was cooking meat. Once the hair charred and curled, we took our knives and scraped it off the hide. We did this again and again until we were sure it was properly cleaned.

We cut the skin into small cubes and threw them into a pot of water. For good measure, Charity even made me sneak into my mother's kitchen and steal a handful of soda.

"It's how women make their beans cook faster," he said. "I'm thinking it works with skin, too."

We let the skin boil for over two hours, adding more water, salt, and soda. After three hours, a thick white foam had collected on the top. Charity took his knife and fished through the froth, pulling out a piece of steaming hide. It was gray and slimy. Charity blew on it for a few seconds, then shoved it into his mouth. He struggled to chew, his jaws working hard, then finally swallowed.

"How is it?" I said, my mouth watering.

"A bit tough," he said. "But we're out of firewood. Let's eat."

We fished the pieces out of the pot and grabbed them with our fingers. The skin was slimy and sticky, as if covered in scalding glue. I put the first piece in my mouth and breathed in, feeling the heat of hot food rush into my stomach and lungs. I chewed and chewed. The juice from the skin seeped out of my mouth and caused my lips to stick together. With each chew, my lips sealed shut.

"Merry Christmas," I said, struggling to speak.

Just then, I heard a clawing sound at the door, then a soft whine. I threw open the door and found Khamba. He must've smelled the Christmas meat all the way from my room and came limping over. His bony frame was hunched and tired, but his tail was wagging as fast as ever. I was glad to see him.

"Give some to that dog," Charity shouted. "It's dog food we're eating anyway."

"For sure," I answered, then turned to Khamba. "Let's get you something to eat, chap. I'm sure you're starving."

I tossed up a piece of slimy skin, and to my surprise, Khamba leaped onto his hind legs and caught it in the air. Just like old times.

"Good boy!" I shouted.

It took him one second to swallow it whole, then he licked his lips and waited for more. I went back inside and brought out two giant handfuls of hide. After slurping up every piece, the life seemed to return to his body. He blinked his eyes excitedly and flipped his tail. And making an exception on Christmas, Charity even let Khamba come inside the clubhouse.

I lost count of how many pieces I ate myself. But after about half an hour of chewing, Charity and I gave up. Our jaws were too tired and sore to continue. Several large pieces of skin remained in the pot, and I thought about my sisters and parents who were at home, probably hungry and dreaming of meat on this Christmas Day. But I didn't dare ask Charity to allow me to share. It was a well-known rule that whatever happened in *mphala* stayed in *mphala*. We'd eat the pieces ourselves the next day.

As the sun went down that afternoon, we sat around a dead and smoldering fire, content with the warm feeling of meat in our stomachs, because that's what Christmas was all about anyway.

CHAPTER SEVEN

*T*HE NEXT WEEK, I received information better than any Christmas gift. I was sitting at home listening to the radio when I heard a wonderful announcement.

"The National Examinations Board has released the results of this year's Standard Eight exams," the announcer said. "If you'd like to check your score, the Board asks you to visit the institution where you sat for your test."

"My scores," I said to my mother. "My scores are ready!"

I raced down the trail toward Wimbe Primary, leaping over the puddles and pits in the trail, feeling such confidence that my head began swimming with the great possibilities. Which boarding school would I be going to? Chayamba or Kasungu? Since I'd decided I wanted to become a scientist—not just *any* scientist, but a great one—I knew those two schools had the best teachers, along with libraries and laboratories and everything I needed to achieve that dream. I didn't care which one I was chosen for. Wherever those chaps needed me, I'd happily go.

The list was posted on the wall outside the administration building. A few other students were also there. I pushed through them, a busy, determined man. The various schools were posted with students' names listed below. I quickly found Kasungu and scanned the names beneath it. Nothing. Moving toward Chayamba, my finger slid over the names Kalambo, Kalimbu . . . then Makalani.

Wait a minute, I thought. *There must be some mistake . . .*

I scanned the lists again, but my name wasn't there.

"Here you are, Kamkwamba," said the boy behind me. His name was Michael, one of the top students. "You're going to Kachokolo."

Sure enough, my name was beneath Kachokolo Secondary School, possibly the worst one in the district. Kachokolo was a community school, or a village school, and not a top priority when the government sent their funding. *How could this be?* I thought.

The exam grades were posted on the next board. Finding my name, I soon discovered why:

MATHEMATICS: C
PRIMARY SCIENCE: C
ENGLISH: C
CHICHEWA: B
SOCIAL STUDIES: D

My heart sank. I imagined walking the long road to Kachokolo, which was about five kilometers away. The school was near one of the big tobacco estates and that road was usually filled with mud and insects. A big dam was nearby, and sometimes Gilbert, Geoffrey, and I went fishing there.

"Congratulations," said Michael, laughing. "You've been chosen to go to the dam school. If anything, you'll become a great fisherman!"

"In two years I can take my Junior Certificate Exam," I answered, thinking aloud. "Then I can transfer to a better school. You'll see me soon in Kasungu. Don't laugh at me now."

"Good luck," he said, laughing anyway.

That's what I would do, I decided. I'd study and become the best student at this village school, then take my JCE exam and impress them all. At Kasungu and Chayamba, they'd be on their knees begging me. In the meantime, there was one good thing about going to Kachokolo. Gilbert had also been assigned there (his grades had also stunk). I thought about the two of us walking to school together, and that made me feel more excited. I had two weeks to prepare.

The new year arrived with daily rains that watered our maize seedlings and encouraged them to grow. By now their stalks were deep green and reached my father's shins. The rains had also allowed us to transport our tobacco plants from the *dambo* to the fields, where the seedlings were now growing strong.

The rains also brought the forests alive, caused the flowers to bloom and the bushes to blossom. Everywhere you went, the land was rich and green, and the air smelled sweet and fragrant. It was a spectacular joke, of course, because nothing was ready to eat.

The rainy season was when the insects bred and hatched their young, when blowflies swarmed the latrines, coming up from the septic hole fat and greenish black. Outside, the flies were so thick you couldn't escape them. They collected on your legs and feet when you were standing still, and on your face when you were trying to talk. People were swatting flies everywhere they went.

There were also great clouds of mosquitoes, carrying malaria and death between their wings. The roaches were also more abundant and bigger than usual. At night after a heavy rain, the valley was alive with the sound of millions of frogs: some that whooped and gurgled, while the call of others was like giant drops of water falling from heaven: *bloop . . . bloop . . . bloop.* The rainy season was for mud and insects, a time when only the geckos and spiders got fat.

For the people, the rains merely brought further misery to those struggling to survive. Even though no one could afford to pay the *waganyu* to work, they continued to walk the roads with their bundles and hoes, stepping through the mud while the rain soaked them from above.

By now, the cost of maize had reached nearly a thousand kwacha a pail, and the *waganyu* and most everyone else had turned to a strict diet of *gaga*. Once the *gaga* began to run low, the traders started mixing it with sawdust. The yellow flakes were somehow concealed in the brown chaff, never revealing themselves until the porridge soured people's stomachs. Once this was discovered, many gathered around the traders who were still selling it in the market.

"I spent all my money, only to get a bellyful of sawdust!" they shouted.

"My children are home sick!"

"This isn't human!"

They could complain, but with no other money in their pockets, there was little they could do.

AT HOME, MY MOTHER spent hours each day baking her cakes, mixing the batter and placing the drops into a metal pot, which she then buried in the coals. She did this over and over until she had a hundred cakes, which she placed in a large basin covered in cloth. She then filled a pitcher with water and took a few cups, so the starving ones could finish filling their stomachs with liquid after eating the cakes. Each evening when my mother came home, we grabbed the basin from her hands, pretending to help her, only to peek under the cloth to see how much flour she'd managed to earn.

Most of her customers were still farmers who'd either sold their worldly possessions, or else had taken loans from the traders and businessmen, who were now charging 300 percent interest. Those were the terms, and if you didn't like them, perhaps you could interest them in your dishes or the roof off your house. Most people took the loans, either because they had nothing to sell in the first place or their homes were already empty.

With these conditions, the situation in the trading center became more and more intense. People would gather around the traders like Mister Mangochi and complain. But these complaints rarely came with threats. People had no energy for violence.

"This is our food," they'd shout. "Why are you starving us with these prices?"

"The farmers in Tanzania are charging us double," Mister Mangochi would answer, telling the truth. "If I lower my prices, I'll be finished and you'll have nothing tomorrow."

One afternoon my mother arrived at the market as usual, set up her table, and within seconds, a mob descended on her stand, shouting and grabbing at her things.

"I'll take two!" one woman screamed.

"Give me three!" said a man.

With this sudden rush and chaos, my mother didn't notice that a few people were snatching cakes from the basin. Others placed an order, but then grabbed what they wanted and fled. One man even sat down next to my mother and said, "I'll take three." He grabbed three cakes from the basin and ate them quickly.

"Nine kwacha," my mother said.

"I don't have any money," he answered.

That evening when my mother came home, her hair was wild and her face drawn and tired.

"Mama, you must be exhausted," I said, grabbing the basin to have a peek. It was nearly empty.

"They took almost everything," she said. "There won't be much tonight."

She was right. That night we had a very small supper.

As the price of maize continued to rise, my parents had to cut back even more on the amount they purchased. This meant that my mother had a smaller number of cakes to sell, which meant our family had less to eat. Our blob of *nsima* began to shrink, to five mouthfuls, then four mouthfuls.

"Every time you put *nsima* into your mouth, add some water," my mother instructed. "That way you trick your stomach."

Although most of us kids tried to be conscious of our portions, my sister Rose, who was seven years old, would grab large handfuls of *nsima* and stuff them in her mouth before anyone could stop her.

"Hey, slow down," shouted Doris, who was two years older. "Mama, she's taking bigger pieces!"

"Maybe you should eat faster," snapped Rose.

We were all becoming thin, especially Rose and Mayless, the two youngest above Tiyamike. They unfortunately took after my mother—small and lithe—so the hunger ate at them more and revealed itself more clearly in their sunken faces. Aisha, Doris, and I were taller like my father, so we had a bit more to lose, although by then I'd already fashioned a belt from a long cloth to keep my trousers up. I had to get more creative as the weeks went on.

My parents never scolded Rose for taking more than her share. But Doris soon reached her breaking point. Over the past weeks she'd become paranoid, fearing she wouldn't get any food at supper and my parents wouldn't help her. As a result, meals became moments of stress and anxiety.

One evening as we sat around the bowl of *nsima,* Rose reached in and pulled off a large piece as usual. But before she could shove it in her mouth, Doris leaped across the basin and began punching her in the face.

"Mama!" screamed Rose.

"Stop it!" my mother shouted and pulled them apart, a task that seemed to drain the rest of her energy. "Just try to eat and get along. I don't have the strength to deal with you two."

That night we went to bed hungry again, the smell of food still strong on our fingers, a smell even water couldn't wash away.

As FOOD RAN OUT across Malawi, the government remained silent. Every day we'd listen to the radio for news about the hunger, but nothing was ever said. With the silos of Press Agriculture empty and their own workers begging for food, and not a kernel more of maize at ADMARC, there seemed to be no rescue on the way. Hunger began to breed paranoia, and with it, rumors spread.

"They sold all of our maize!" people said. "What else have they sold?"

"Nothing is safe in Malawi!"

"What are they trying to hide from us?!"

Convinced the government was collapsing, farmers across the district rushed to the banks in Kasungu to withdraw the little savings they had left. One morning in a heavy rain, my father caught a pickup with several others and rode into town. When he arrived, the line stretched out the door and down the street. About a hundred men waited all day in the rain with nothing in their stomachs except anger and fear. Once they got inside, the overwhelmed clerks told them to wait even longer. With this, the farmers threatened to riot.

"Give us our money now!" they shouted. "What are you trying to hide from us?"

My father managed to withdraw our family's entire savings—about a thousand kwacha—and used it to buy another pail of maize, which he milled and sold the next day. We'd eat another week.

DESPITE WHAT WAS HAPPENING, I had something to look forward to in mid-January when I started classes at Kachokolo Secondary School. For weeks I'd focused on this one day, how I'd finally wear my long trousers, and Gilbert and I would walk like big men into our future. Thinking about school had kept my mind from dwelling on the troubles. Somehow, being hungry at school seemed easier than being hungry at home.

The only problem was my uniform. I had a pair of black trousers, but my parents couldn't afford to buy a proper white shirt from the headmaster. Instead, my mother sent me to the used clothing stalls in the trading center.

"What difference does it make where it comes from," she'd said, "as long as it's white?"

Well, I had only two shirts in total, and even before school started, I'd been forced to wear my uniform shirt and had gotten it dirty. Then we ran out of soap. We'd managed to buy one tablet of cheap Maluwa lye soap at the first of the month, but it had already vanished. We could wash our bodies with warm water and *bongowe* bushes, but a white shirt wasn't so easy.

One morning I pulled out the half tractor tire we used as a washing station and filled it with hot water, letting the shirt soak until the water was cool. I then scrubbed it with *bongowe,* but it didn't work. Yellow circles remained around the underarms and my collar was still gray. What could I do?

On the first day of class, I met Gilbert on the road so we could walk together.

"Gilbert, bo?"

"Bo!"

"Sure?"

"Sure!"

"My friend, this is the day we've been waiting for!"

"It is indeed!"

"Are you ready for the teasing and bullying?"

"Yah, who do you think will hit us first?"

"Well, I've been thinking of a plan. If a Form Three boy approaches us, and he's not quite muscular, we should deal with him straightaway."

"Oh, that's a good plan! That will show everyone we can fight."

"Exactly."

"So who will hit him first?"

"I think you should."

THE FORTY-MINUTE WALK TO Kachokolo took us over hills and across maize fields and past the *dambo* marsh where we once used to hunt. The school sat in a valley surrounded by tobacco estates, where I could see giant diesel tractors tilling the soil and the few lucky men with jobs working in the sun.

Once at school, we gathered in the yard surrounded by blue gums for our first morning assembly. The headmaster, Mister W. M. Phiri (no relation to the legendary magic fighter), stood up and told us how happy he was to see such bright, promising new faces.

"Here at Kachokolo," he said, "you'll be given the knowledge to help your country and make it proud."

We certainly were a fine bunch, all of us eager to learn and squirming with excitement. At that moment, I was certain I was experiencing the greatest day of my life. I couldn't stop smiling.

"But just like in any institution of learning," he continued, "this school has rules that must be followed. Every student should be in a proper uniform and be punctual. If not, the punishment shall be swift."

When assembly was over, I was walking to my classroom with Gilbert, when Mister Phiri tapped me on the shoulder.

"What's your name?" he said.

"William Trywell Kamkwamba," I said, unable to hide my nerves.

"Well, William," he said, "this is not the proper uniform."

He must've noticed my underarms. I wanted to run and hide. But Mister Phiri was pointing down at my feet.

"Sandals aren't allowed," he said. "We require students to wear proper footwear at all times, so please go home and change."

I looked down at my flip-flops, which had certainly seen better days. The rubber connecting the sole had broken on one foot, forcing me to carry in my pocket a crochet needle and bit of rope for emergency repairs. I had no shoes at home. I had to think quickly.

"Mister headmaster, sir," I said. "I would put on proper footwear, but since I live in Wimbe, I must cross two streams to get here. And because it's the rainy season, you can imagine how the mud will destroy my good leather shoes. My mother wouldn't have it."

He scrunched his eyebrows and considered this. I prayed it would work.

"Okay," he said. "It's fine for now, but once the rains are over, I want to see you in *proper* footwear."

My parents had no money for schoolbooks, either, which cost hundreds of kwacha each. Even in better times, most students couldn't afford these books and were forced to share. At least in primary school, that meant squeezing your bottoms together on the same chair and hoping the other person didn't read faster than you did. But luckily for me, Gilbert had managed to buy his own books and said I could look on, which was good because we read at the same level.

At Wimbe Primary, the conditions had been dreadful, with students having to read and study outside under trees because the classrooms were too full. But even inside the classrooms, the roofs leaked when it rained. The Standard Three classroom was missing an entire wall, and the latrines were not only disgusting but dangerous. Termites had eaten away the floor planks, and one afternoon, one of my classmates named Angela actually fell through. It took hours before anyone heard her screams from the slimy bottom. We never saw her at school again.

I'd been hoping for better conditions at secondary school, but no such luck. Once Gilbert and I arrived in our classroom at Kachokolo, our new teacher, Mister Tembo, told everyone to sit on the floor. It seemed the government had sent no money for desks, and from the looks of things, they hadn't sent money for repairs, either. A giant hole was in the center of the floor, as if a bomb had exploded.

Right away we began studying history, covering the early civilizations in China, Egypt, and Mesopotamia. We learned about written history and oral history and the early forms of writing. We started algebra in math, which I found incredibly difficult. We also started learning a bit of geometry, and this I absolutely loved. In geometry, we learned about angles and degrees, and I remembered builders in the trading center using such terms.

One afternoon in geography class, Mister Tembo pulled out a world map, and we located the continent of Africa.

"Can anyone find Malawi?" he asked.

"Yah, here it is!"

We ran our fingers over our country, and I marveled at how small a place it was compared to the rest of the earth. To think, my whole life and everything in it had taken place inside this little strip. Looking at it on the map—shaded green with roads zigzagging brown, the lake like a sparkling jewel—you'd never guess that eleven million people lived there, and at that very moment, most of them were slowly starving.

DESPITE WHAT I'D IMAGINED earlier, the hunger was just as painful in class as it was in the fields. Actually, it was worse. Sitting there, my stomach screamed and threatened, twisted in knots, and gave my brain no peace at all. And I soon found it difficult to pay attention. During the first week of school, enthusiasm among my fellow mates had been high, but only two weeks later, the hunger had whittled away at all of us. A gradual silence soon fell over the entire school. At the beginning of term, a dozen hands would shoot up when Mister Tembo asked, "Okay, any questions?" Now no one volunteered. Most just wanted to go home to look for food. I noticed faces getting thinner, then some faces disappearing altogether. And with no lotion or soap at home, people's skin became gray and dried, as if they were covered in ash. At recess, talk of soccer was replaced by tales of hunger.

"I saw people yesterday eating the maize stalks," one boy said. "They're not even sweet yet. I'm sure they became sick."

"I'll see you gents later," said another. "I'm not coming back tomorrow. I don't think I can manage the walk anymore."

I suppose none of this really mattered because on the first day of February, W. M. Phiri made this announcement at assembly. "We're all aware of the problems across the country, which we also face," he said. "But many of you still haven't paid your school fees for this term. Starting tomorrow, the grace period is over."

My stomach tied itself into another knot because I knew my father hadn't paid my fees. I'd refused to ask him about it these past weeks, knowing what he'd say. The fees were twelve hundred kwacha, and that was collected three times a year. Walking home, I cursed myself for being so optimistic, for allowing myself to become so excited. I wondered why my parents had even allowed me to go to this school in the first place.

"What am I going to do?" I asked Gilbert. "I suppose I'm going home now to face the music."

"Don't stress too much, *eh*?" he said. "Just wait and see what happens."

When I got home, I found my father in the fields.

"At school they're saying I should bring my fees tomorrow, twelve hundred kwacha," I said. "So we should pay them. Mister Phiri wasn't joking around."

My father looked down at the dirt the same way he'd looked at those sacks of grain in the storage room—as if waiting for it to tell him something. He then gave me the look I'd grown to fear.

"You know our problems here, son," he said. "We have nothing."

I was watching his lips move, but in my head, I heard Mister Phiri's voice, repeating what he'd said as we walked out of school: "No fees. No school."

"I'm sorry," said my father. "Next year will be better. Please don't worry."

I could see that my father felt terrible, but I was certain it was nothing compared to how sad I felt. The next morning, perhaps just to torture myself, I woke up at the same time, stood at the junction, and waited for Gilbert. I even wore my black trousers and white shirt, though I don't know why. I wasn't going anyplace.

Soon Gilbert appeared and walked past. "Come on, William," he said. "Aren't you coming to school?"

I wanted to cry but didn't. "I'm dropping out," I said. "They don't have the fees."

"Oh, sorry, William," he said. Gilbert looked quite upset, which somehow made me feel better. "Perhaps your parents can find the money."

"Perhaps," I said. "I'll see you later, Gilbert."

Feeling terrible, I walked over to Geoffrey's house to share my burden. A few weeks before, Geoffrey had a nice stroke of luck. During a big storm one night, a bolt of lightning had struck a giant blue gum behind his house and caused it to fall. The next morning Geoffrey was outside hacking it apart with his panga. He'd wandered the roads selling clumps of firewood for thirty kwacha apiece, which had allowed his family to eat for a couple of weeks. Between my school and his firewood, I hadn't seen him much. I was anxious to catch up.

Geoffrey was getting dressed when I walked in, and the very sight of him caused me to stop in place. His clothes looked borrowed, as if they belonged to someone else. His eye sockets were sunken and dark. The whites of his eyes appeared very white, as if he were a ghost or a spirit. My cousin had once been a big guy, a real bouncer, but now he was down to nothing. It had happened so fast.

"Why aren't you in school?" he asked. "Didn't you get selected for Kachokolo?"

"No money," I said. "Today I dropped."

"Oh . . . sorry to hear. Me and you, we're in the same hopeless situation."

"For sure."

He stared down at the floor and shook his head.

"Hopefully God has a plan for us."

Even after seeing Geoffrey's condition and leaving his house, I still sulked around feeling sorry for myself, thinking, *eh, why me?* as if I was the only one stricken with such luck. But that afternoon when school let out, Gilbert stopped by again and told me the news.

"Today we were few," he said. "Most of school dropped out as well."

Out of seventy students, only twenty remained.

My own problems didn't seem so important; the hunger belonged to the entire country. I decided to put faith in my father's word, that once we made it through the hunger, everything would be okay. But first we had to make it through the hunger. And as Geoffrey had said, it was hard enough just worrying about tomorrow.

By late January, the *GAGA* was finally gone. The people who'd depended on this for *nsima* turned instead to pumpkin leaves, and here the real starvation began. Famine arrived in Malawi.

It fell upon us like the great plagues of Egypt I'd read about, swiftly and without rest. As if overnight, people's bodies began changing into horrible shapes. They were now scattered across the land by the thousands, scavenging the soil like animals. Far from home and away from their families, they began to die.

The same people I'd seen carrying their belongings to the trading center now stumbled past us in a daze, their eyes swimming in their sockets.

The hunger ravaged the body in two different ways. Some people wasted away until they looked like walking skeletons. Their necks became long and thin like the *dokowe* birds that drank near the river, not even strong enough to hold up their heads. Others were stricken with kwashiorkor, a dreadful condition a body gets when there are no proteins in the blood. Even as these people starved, their bellies, feet, and faces swelled with fluid like ticks filling with blood.

The starving people didn't say much as they passed. It was as if they were already dead, yet still looking to fill their stomachs. They moved carefully along the roads and through the fields, picking up banana peels and discarded cobs and stuffing them into their mouths. Near my house, groups of men were digging up the *chikhawo* roots of banana trees so they could boil them like cassava. Some dug up other roots and tubers, even the grass from the roadside, and milled them into flour. Others resorted to eating the seeds from government starter packs, scrubbing off the pink and green insecticide that kept off the weevils. But it was impossible to get all the poison off, and many suffered from vomiting and diarrhea, which only made them weaker. Plus, having now eaten their seeds, they had nothing left to plant.

Every day the starving people stopped by our house and begged my father for help. I'd see them coming from far away and think, "Oh, this chap is looking fat. Finally someone who is doing well." But once they arrived, I realized it was only the swelling.

The starving people saw we had iron sheets on the roof of one building and thought we were wealthy, even though our sheets were only fastened by stones. Some of the men had walked twenty and thirty miles.

"If you have one biscuit, please, I can work," they'd say, their bare feet too swollen to wear sandals. "We're now six days without eating. If you have just a small plate of *nsima*?"

"I have nothing," my father had to say, "I'm barely feeding my family on one walkman."

"Just give us porridge," they demanded.

"I said no."

A few men still chose to stay in our yard all night. The ground and wood were too wet for fires. When the rains came in the darkness, the men curled up under our porch and shivered. By morning they were gone.

A few nights later, we were seated outside having our meal when a man approached from the road. He was covered in mud and so thin it was hard to understand how he was even alive. His teeth poked out of his mouth and his hair was falling out. Without saying hello, he simply sat down beside us. Then, to my horror, he reached his dirty hands into our blob of *nsima* and ripped off a giant piece. All of us sat there in shock, saying nothing as he closed his eyes and chewed. He swallowed long and satisfyingly. When the food was safely in his stomach, he turned to my father.

"Do you have more?"

"I'm sorry," my father said.

"Okay then," the man answered, then stood up and walked away.

WILLIAM KAMKWAMBA AND BRYAN MEALER

The Wimbe trading center, just near the primary school. The trading center became a kind of ghost town during the famine, despite being full of starving people.

THE CROWDS CONTINUED TO pour in from the bush. More than ever, they now converged at the trading center like herds of crazed and wasting animals driven together by fire. Women with thin, ashen faces sat alone, pleading with God. But they did it quietly and without tears. Everywhere the anguish was silent because no one had the energy to cry. Elsewhere in the trading center, children with swollen stomachs and strange copper hair clustered under storefronts. A few traders still spread tarps in the mud and sold grain, but the units had become smaller and smaller. Its price was like gold, like buying the universe and stars in one half kilo. Crowds gathered round, but mostly to stare in stunned silence, as if watching a dream in heaven. Those with energy still screamed and begged.

"*Bwana*, just a small plate of flour," they said. "That's all I need. It's for my child."

"*Eh,* if I start with you, then . . ." the traders answered, then said nothing at all.

Those with energy still stayed close, lunging like dogs whenever a kernel fell to the ground, scooping gravel and all into their mouths. One man weaved through the crowd mumbling, "Please help, I'm an orphan. My father has died." The man was forty years old.

Each group told a different story of the dead.

"I heard there was a man who'd spent days looking for food," one farmer said. "One morning he decided to nap under a tree and never woke up."

"I was cooking walkman," said another, "when a man sat down uninvited, telling me, 'I need to eat this.' But before the *nsima* was even ready, that man was dead."

Others had gone so many days without eating, the first bellyful of food had sent their bodies into fatal shock. One woman stepped over two dead men on the road, still clutching their hoes. Men who'd swelled with kwashiorkor tried to relieve their suffering by draining their great blisters with a knife, only to die several days later from infection.

News came of Beni Beni, the madman of Wimbe, who'd always made us laugh in better times. He'd run up to merchants in the trading center with his raving eyes and snatch cakes and Fantas from their stalls. No one ever took them away because his hands were always so filthy. The mad people had always depended on others to care for them, but now there were none. Beni Beni died at the church.

Amid this great suffering and confusion, the government radio said the president had traveled to London on state business. When he returned, a Malawian radio reporter questioned him about the famine. We all gathered around at home to listen.

The reporter said something like, "Your Excellency, many are dying across the country from lack of food. What are you planning to do?"

The president scoffed at this absurd idea, saying he'd grown up himself in the village where people often died of other things, such as tuberculosis, cholera, malaria, or diarrhea, but never from lack of food.

"Nobody has died of hunger," he said.

When the report was over, my father shook his head and turned away.

"How could he say such a thing, Papa?" I asked.

"Some men are blind," my father replied. "But this one just chooses not to see."

That afternoon, the ways of the world suddenly became more clear. Whereas I was still confused as to how the hunger had been allowed to happen, this much was certain: Every man for himself. We were on our own.

CHAPTER EIGHT

*N*OT LONG AFTER THE radio report, my mother came home again with very little flour for supper, and that night all we had was a taste. As I sat on the floor to eat, I looked down the corridor and saw Khamba standing at the open door. He was looking bad, with his head hung low and his eyes drooping. His ribs were like blades against his skin. The walk across the courtyard exhausted him.

He was starving to death.

The last big meal he'd eaten was our Christmas goat skin. That food had given him a bit of strength and even added some weight to his frame. But there'd been little since. I counted how many times I'd fed Khamba in the past two weeks—just handfuls of *nsima*, nothing more—and stopped at five. I didn't have to think very hard. That number was all I ever thought about, and seeing him standing there was like a hammer in my stomach. I had nothing to give him tonight.

"Sorry, friend," I said. "I just can't share."

It didn't take me long to eat. When the food was gone and out of sight, I stood up and made my way down the corridor. Stepping over the dog, I walked back to my room, closed the door, and got in bed.

The next morning, the hunger woke me up. Little did I know, but my stomach had taken over my entire body, filled every limb and crevice, all the way up to my head like a great big balloon. At some point in the early morning, it had finally burst and revealed its emptiness. It had only been filled with air, and in this nothingness, there was only pain. I took deep breaths to try and fill the space again, but it was no use. I was flattened like a tube. It hurt so badly. I lay in bed and listened to the rain pound my ceiling steadily through the grass thatch and plastic sheeting below. Somewhere in the dark, it dripped and dripped.

I have to eat, I thought.

I lay there until the sun sent a gray beam through my shuttered window. When I no longer heard rain drops hitting the roof, I gathered the strength to sit up. I got out of bed and pulled on my clothes, grabbed a few things, then made my way outside. I stopped at the door of the kitchen and peeked in. Khamba was curled up near the fire pit, which had long lost its heat. I couldn't tell if he was breathing. I wondered if we'd had the same dream.

"Khamba!" I shouted. "Let's go hunting!"

The very word was like a bolt of current through his body. Khamba jerked his head up and slapped his tail once against the floor. It had been a year since I'd said those words. Despite his weakness, he struggled to his feet. His legs trembled from arthritis, but his tail swung like mad. He was ready.

"Let's get some food!"

I had no maize or *gaga* to set my trap, so I grabbed a handful of ash from my mother's fire and threw it into a sugar bag. The dog and I then walked outside toward the Dowa Highlands, which seemed forever draped in turmoil. The rains would never end. The walk took twice as long, with Khamba limping slowly behind. I stayed just a few steps ahead, whistling a tune to try to raise my spirits. The maize in the fields was growing tall and green, and I knew it wouldn't be long until we ended this torment. Maybe a month, perhaps a bit longer. I scanned the sky for birds but saw none.

At the trap site, I fastened my rubber to the poles, stretched it back, and set the trigger. I sprinkled the ash onto the kill zone. It looked pathetic.

"Let's hope this works," I said.

If we could get three birds, perhaps tonight I could sleep. Perhaps Khamba could get better and hold on—just for another month. I took the rope and dragged it twenty yards behind the *thombozi* tree. Khamba collapsed with a thud, then nearly fell asleep. I stretched out on my stomach and waited.

About fifteen minutes later, a small flock of about five birds swooped down and landed outside the trap. Khamba's head shot up, as if he'd seen it in a dream. As the birds were nearing the kill zone, my imagination began to wander:

In slow motion, I saw the rubber smash the birds against the bricks, crushing them instantly. I saw myself plucking the feathers. I pulled off the heads, took my knife from my waist and made an incision just below the breast, then scooped out the entrails. I tossed the bloody bits into the air, and Khamba snatched them in his jaws. I laughed and praised him, but I couldn't make out the words. After that, I rinsed my hands with water from the *dambo,* getting off all the blood, then rinsed the birds clean. I then sprinkled the warm flesh with salt and rubbed it into the skin so it was sure to stay. I started a small fire behind the tree, which quickly died to a glowing bed of embers. I split the backsides and laid the birds' breasts directly on the coals and heard the meat begin to sizzle. Soon the smell of roasting flesh was all around me, encompassing the entire universe. It tickled the back of my brain. I flipped them over, salivating.

The sudden pounding of my heart snapped me out of the spell, just in time to see the birds hop toward the trap. Khamba sat rigid and watching, lost in his own delicious daydreams. I gripped the rope, muscles twitching. But as soon as the birds stepped into the kill zone, they realized the bait was only ash and quickly flew away in a burst of wings. I exhaled in defeat and dropped the vine, too exhausted to move. I may have even cried.

THAT NIGHT AT HOME, Khamba curled up outside my door and fell into a deep, almost frightening sleep. The hunt had taken nearly everything from him. At supper, I saved half a handful of *nsima,* plus some pumpkin leaves, and walked outside to where he lay.

"Khamba!" I shouted. "Supper!"

It was another word he knew. He opened his eyes and slapped his tail. I tossed the food above his head, but he made no effort. The *nsima* and pumpkin leaves hit the ground with a thud. Almost as an afterthought, Khamba picked himself up and ate.

Two days later, I fed him again. It was only a bit of pumpkin leaves, so I walked out and put them in his food bowl. He saw me and limped over. But as soon as the pumpkin leaves hit his stomach, he vomited them right up. I knew the end was near.

"Just wait one month," I said, pleading. "We'll be feasting in a month!"

The following evening, he vomited his food again. Nothing worked.

The next morning Charity and Mizeck walked over to my house on their way to the trading center. I hadn't seen Mizeck since we'd feasted on my birds at the clubhouse. He'd once been a fat guy, now there was hardly anything left. I could see the skull in his face. When he saw Khamba, his voice got a bit crazy.

"Look at this thing," he said, standing over Khamba. "He looks pathetic!"

Khamba was half asleep, his skin and bones now covered in flies. He'd stopped even caring.

"I can't even stand to look at it," Mizeck said.

Then don't, I thought, but said nothing. What did he care about my dog? I tried to change the subject.

"So what are you chaps doing today?" I asked.

"Going to the trading center, as usual," said Charity. "Looking for some *ganyu,* but I'm not so hopeful."

As Charity and I chatted, Mizeck remained quiet. He was still staring at Khamba, not even blinking his eyes.

"Why don't you put this thing out of its misery?" he said. "Take it behind the house with a big stone."

I pretended not to hear.

"He's right, William," Charity said. "You need to do something. Take him to the *dambo*. The water is high. This dog is too weak to swim."

"Wait a minute, guys," I said. "What are you saying?"

"We're saying it's time to be a man," said Mizeck. "It's time to kill this pathetic animal."

I wanted to smash his face. "Guys, Khamba's fine," I said, my voice growing weaker.

"Listen," said Mizeck, "if you can't be a man and do it yourself, we'll do it for you. This thing is making me sick."

"It's the right thing to do," said Charity, lowering his voice. "You won't even have to bother. We'll stop by tomorrow and take him. He won't feel a thing."

I struggled for words, some form of protest, but was stopped short. Mizeck now glared at me, his eyes wild.

"It's not your decision," he said.

When Mizeck and Charity left, I felt dizzy and weak, as if my legs were made of grass. I stood there watching Khamba sleep, then managed to sit down beside him. The flies were thick on his coat, circling and landing, circling and landing. After half an hour, he finally opened his eyes and caught me staring, then gave me a slow slap of the tail. The way he looked at me, even like this, made me remember our days hunting, how we could talk to each other without even speaking.

How would I protect my dog if Mizeck and Charity returned? I couldn't let them take him. I thought about the many ways of doing this until the compound grew dark, then finally realized my answer. They'd been right. Khamba was suffering, even miserable. But they'd been wrong about me.

The next morning, I was back outside watching Khamba sleep when Charity appeared in the courtyard. My heart began to pound. He looked at my dog and cocked his chin. But before he could say anything, I stood up.

"I'm taking him," I said.

"What?"

"I'm taking him to the forest."

He shrugged. "A stone is very quick," he said.

"This is what I want."

Charity nodded his head. "You're making the right decision. We'll do it together, today."

"Today," I said.

That afternoon, when Charity returned, we walked to the shady area outside my room. Khamba still lay there, unmoved. I then pretended I was someone else, and so I was.

"Khamba!" I said. "Let's go hunting."

His head perked up.

"I said let's go!"

He stumbled to his feet, shook once to get the flies off his back, and hobbled toward me. It took us ages to get out of the compound. I walked backwards slightly ahead of Khamba, making sure to never take my eyes off him.

"Come on, boy, you can do it."

We walked down the road toward the highlands. The sun was low in the west and colored the hills in an orange glow. The air was warm and dry, the perfect weather to hunt. We entered the blue gum groves that reached just below our heads, the bush that Khamba knew so well. At one point, Charity turned.

"This way," he said.

My heart dipped into my stomach. Khamba followed, struggling to walk over the high grass and sticks.

"Come on, Khamba," I said.

I could feel the tears hot in my throat, but I swallowed them down. Charity turned to me.

"Don't be so upset," he said. "It's just a dog."

"Yah," I said. "Just a dog."

After several minutes, we stopped in the thick brush, surrounded in chest-high grass. The mountains were visible in the distance through the blue gums.

"This is a good place," Charity said. "No one will pass here."

I looked around. I could still see my house. "This is too close!" I said.

"This dog can't go any farther."

Khamba had already flopped down under a *thombozi* tree and was panting heavily.

Without my saying anything, Charity began stripping the bark off a few *sanga* trees to make one long rope, then doubled them together to make it strong. I turned my back and stared into the trees.

At one point Charity's hands went quiet, but I didn't look.

"Tie him to the tree," I said.

Charity lashed the rope around the trunk of the *thombozi* tree, then tied the other end around Khamba's front leg.

When I got the courage to turn around, I saw my dog lying in the tall, bent grass. His ribs jutted out his sides. He was panting, weak. Without saying anything, Charity turned and walked away. When I followed, Khamba lifted his head and began to cry. His body was too weak to make any real noise, just a panicked whine that came from deep within. He knew I was leaving him. After a few steps down the trail, I made the mistake of turning back around. His eyes were still on me. Then he laid down his head.

"I did a terrible thing," I whispered, walking faster. I was going to vomit.

"He was old," Charity said. "He was going to die anyway."

"I did a terrible thing."

Once we reached the compound, Charity went to the clubhouse, and I walked toward my room. On the way, my eyes caught sight of Khamba's food bowl lying by the henhouse. I ran over, picked it up, and hurled it against the ground, shattering it into pieces.

It's just a dog, I thought.

That night I stayed awake long into the night, knowing that Khamba was just down the hill. If I screamed his name loud enough into the dark, he would probably hear me.

The next day I avoided most people and tried to stay in the fields. But once I got home, my uncle Socrates was coming out of my house, having visited my father.

"Where's Khamba?" he said. "I can't find him anywhere."

"I haven't seen him," I said. "I was wondering the same thing."

"Hmm," he said. "I hope those wild dogs didn't get him. I'm sure we'll smell him soon enough if they did."

I felt nauseous all day. That night I tried to force Khamba out of my mind, thinking of anything but him. It wasn't difficult. I was so hungry that I couldn't concentrate on one thing too long anyway.

The next morning, Charity stopped by and found me in my room.

"Let's go see what happened to Khamba," he said. He was in good spirits.

"What do you mean?"

"Let's go see if he's dead."

I said nothing.

"We'll bring some hoes so people think we're going to the fields," he said. "That way we can also bury him."

We set off down the trail carrying hoes. I was too mixed up inside for conversation. We turned off the road and into the bush. The grass was still wet with dew and soaked my trousers. After some minutes, I saw a white hump on the ground. We continued.

"Is he dead?" Charity asked.

Getting closer, I had a clear view. Khamba was lying in the same position as I'd left him. His head was down, resting on his front paws, and his eyes were wide open. I gasped, expecting him to move. But as I got closer, I saw his tongue was sticking out. It was dry, like paper. A trail of ants moved in and out of his open mouth.

"Khamba is dead," I said.

The rope hadn't budged. There'd been no struggle. A terrible thought suddenly occurred to me: when Khamba saw me leave, he'd given up his will to live. That meant I killed him.

While Charity untied the rope from the tree, I dug a pit with my hoe, my mind blank and black inside. A vigorous energy came rushing upward from that cold, dark place. It was the hardest I'd worked in months.

The rope was still attached to Khamba's leg, so I pulled while Charity pushed with the hoe. With a bit of effort, Khamba toppled into the shallow pit.

"So long, Khamba," I said. "You were a good friend."

We filled the hole with soil and left no marker, even concealed the patch with grass and branches. When Charity and I got home, we told no one about what we'd done. Even after all these years, it's remained a secret, until now.

Two weeks after burying my dog, Cholera swept the district.

The epidemic had started in November in the southern region near Mwanza. A farmer from that area traveled to a funeral in Kasiya, twenty kilometers from Wimbe, and brought the sickness with him. Within days, a dozen were dead in that village, and hundreds were infected across the district.

Cholera is a terrible way to die. It begins with a horrible stomachache, nausea, and instant weakness. Violent diarrhea then follows, milky and colorless and without any smell. It drains the body of all life and energy, leaving a person so weak he can't even speak. Without treatment, death comes in six hours. Across Malawi and the rest of Africa, cholera is an unfortunate companion of the rainy seasons. Many villages have poorly built pit latrines, which sometimes flood and pollute the streams and wells where people drink. The blowflies also carry the bacteria after crawling out of latrines and landing on food. And during the famine, people looking for something to eat also became carriers. The cholera would strike them on the road, forcing them to become sick in the bush. Rain, flies, and cockroaches then spread the disease, contaminating the banana peels, tubers, and husks people picked from the ground to eat.

To keep us safe, the clinic in the trading center began giving out free chlorine to treat our drinking water. My mother brought it home one afternoon in a Coca-Cola bottle, and for the rest of the month, our water tasted like metal. We also covered the latrine hole with a broom handle and flat piece of tin, as advised. But once you stepped inside and pulled off the lid, the flies swarmed from the hole like the great plagues of the Bible, smashing into your face and mouth and around your head. It became quite stressful, swatting them and trying to finish your business at the same time. Any traces of diarrhea around the latrine hole would always cause alarm.

Each day, the cholera people walked past our house on their way for treatment, their eyes milky and skin wrinkled from dehydration. I'd watch them from behind the trees until they got close, then run down the trail toward home. But just as they left, the starving people would follow.

Those who died from cholera were soaked in chlorine and buried at night in the graveyard by the Catholic church, usually by the same doctors and staff who'd treated them. To speed the process, two bodies would be placed in the same shallow hole, then quickly covered. No one knows how many were dying across Wimbe. Between hunger and cholera, there were burials every day.

BACK AT HOME, GEOFFREY'S anemia had grown worse. His legs became horribly swollen, and if you touched his feet, your finger left an impression in his bubbled skin, as if his legs were filled with clay.

"Can you feel it?" I'd ask, poking the blisters.

"I can't," he said.

He was often dizzy and had trouble walking straight lines. One afternoon I took him outside into the sunshine, but he stopped and said, "Wait, I can't see." We had to stand there until his eyes adjusted to the light. For months now, his mother had only served pumpkin leaves at supper. My cousin was slowly dying.

With little else to do, my mother took half our flour for the day, placed it in a basin, and walked over to Geoffrey's house.

"I've come to share this with you," my mother said. "It's not much, but it's enough for porridge."

"Thank you so much," said Geoffrey's mom. "You've saved our very lives."

"We share what we have," my mother said. "We can't allow our family to suffer."

A few days later, my father's sister Chrissy came by and said Grandpa had fainted in his yard.

"He's only eating pumpkin leaves," she said. "Brother, please pray for our father." That afternoon, my mother took half our food again and gave it to Grandpa.

WE WERE ALL LOSING weight. The bones began to show in my chest, and the rope I'd used as a belt no longer sufficed. Now I'd started pinching my two belt loops together, then tying them off with a stick, much like a tourniquet, which I could simply twist when I became any thinner. My mouth was always dry. My arms became thin like blue gum poles and ached all the time. Soon I found it difficult to squeeze my hand into a fist.

One afternoon I was pulling weeds in the fields when my heart began beating so fast that I lost my breath and nearly fainted. *What's wrong with me?* I thought. I was so frightened. I bent down slowly until my knees were in the soil, then I stayed that way for a long time, until my heart returned to normal and I could breathe.

At night I sat in my room with the lamp and stared into the walls, dozing off, passing through some other world. I watched a centipede crawl up the wall for what seemed like hours. I grabbed a mayfly by the wings that had flown too close to the flame, asking it, "How are you alive? What are *you* eating?" then let it go, watching it spiral toward the ground like a broken paper airplane.

No magic could save us now. Starving was a cruel kind of science.

AROUND THIS TIME, MY father started weighing himself obsessively. By now his mighty frame had shrunk like a piece of fruit in the sun. Sharp bones poked through where giant muscles once dominated. His teeth seemed bigger; I noticed his scars. One day he told me he was having trouble seeing across the compound. Like Geoffrey, the hunger had taken his vision.

It seemed the thinner my father became, the more he wanted to weigh himself. He kept a maize scale hanging by a rope from the tobacco shelter, and one morning I watched his routine. He gripped the hook, then hung there like a sack of maize or a bale of tobacco, staring up at the needle. He then made a grunting sound and said, "Hmm, five kilos. Mama—"

As always, my mother came and looked, but she refused to weigh herself. The children were also forbidden. Like many women during the hunger, she'd started tying her *mpango* tight around her waist like a belt. She said it confused her stomach and tricked her heart from beating so fast, helping her to breathe. When she nursed my baby sister in the mornings and afternoons, her hands would shake.

At night, she resorted to mind games.

"You're losing weight because you keep thinking about food," she'd tell my sisters. "Don't you know that causes your body to stress and burn more energy? If hunger is all you think about, you'll only suffer more."

"I don't want to become swollen, Mama," my sisters cried.

"Then think about positive things. Please, do that for me."

She'd then serve our *nsima,* and we'd pass through it like a dream, our three mouthfuls of food leaving no trace in our bodies.

My father then started excusing himself once the food was served.

"Papa, aren't you going to eat?"

"No, I'm fine, you kids go ahead."

One day, sitting in the yard, my father said the strangest thing: "One of the mysterious, yet wonderful things about the hunger is it only kills men."

He sounded mad, but it was true. Men were the ones going out foraging, and in turn, burning precious energy that couldn't be restored. Cholera didn't discriminate, but hunger seemed to take only the men. Lots of men were also abandoning their families and leaving them to fend for themselves. In the mornings, they'd tell their wives they were going to look for food and then never come home. The pressure of providing for a wife and children was just too much to bear. Widows and the forsaken now gathered at the chief's house by the dozens. I hadn't seen Gilbert for days because he was so busy helping care for them.

My father must've been thinking about this, because he turned to my mother and said: "My family is mine to look after. If we're supposed to die, then we die together. These are my principles. God is on my side."

A week later, my sister Mayless became sick with malaria. For days she lay on her reed mat, sweating and shivering in her sleep, then waking up to vomit. She couldn't keep down any food, and like the rest of us, she was already so dangerously thin. At night she'd cry and moan from the pain in her arms and legs. Her fevers became so violent my parents tried admitting her to the clinic, but the doctors refused because it was quarantined for cholera.

At home, my mother stayed up for hours by her bedside, singing traditional songs and soothing her with a wet cloth. A kerosene lamp flickered inside the room.

All night we heard the same words coming through the darkened doorway.

"Don't worry, don't worry."

But everyone was worried.

"Pray for your sister," my mother said. "She's very sick."

When Mayless finally recovered, she was so thin she was like a ghost walking among us.

AROUND MID-FEBRUARY, THE TOBACCO was finally ready to prune, and my father needed me and Geoffrey to help. We gathered the yellow, oily leaves into fisted bundles. Then, sitting in the shade, we threaded the stem of every leaf with a crochet needle and *mlulu* vine. The bundles were hung to dry under shelters made from blue gum and bamboo poles, a process that could take as long as eight weeks, depending on the humidity. Threading and hanging took hours and killed our backs, especially because we had no energy to stand or hold ourselves up. In our delirium, the rows of drying tobacco began to look like delicious food.

"I wish we could eat it," I said to Geoffrey.

"Yah, by now we'd be full."

"Soon it will be dried and the traders will line up to buy it. We'll finally end this sadness."

"Yah, for sure."

But that's not what happened. Only a week after hanging the tobacco, my father went to the trading center and started cutting deals against the crop once it was dry. He couldn't wait for the auction. He had to find our supper.

"Brothers, my family is placing all their hopes on me," my father said. "I'm asking you to give me a quality price, perhaps twenty kwacha a kilo." A full walkman costs thirty.

The traders just shook their heads. "You know times are tough," they said. "I can't see myself giving more than ten kwacha a kilo, not during this time."

"Please, just enough for a walkman," my father said. "Don't make me beg. This is quality tobacco. It's drying very nicely."

After some time, they agreed to fifteen. As the famine deepened, these deals became more sour and unjust.

"I'll give you one bucket of maize for ninety kilos of that tobacco when it's ready," the traders started saying.

Even Mister Mangochi, my father's friend, couldn't resist the market, and my father had no choice but to accept. Each week, he cut more deals against the crop, attempting to keep the numbers straight in his head. Many men would've fought to the death to even have such an opportunity.

Meanwhile, out in the maize fields, the stalks were now as high as my father's chest. The first ears had begun to form, revealing traces of reddish silk on their heads. The deep green leaves had begun their fade to yellow, along with the stem. While men withered and died all around, our plants were looking strong and fat.

"Twenty days," I said, looking at my father.

"I'd say you're right."

We smiled and stroked the leaves like swaddled babes, enjoying the soft music they created together in the breeze.

If I was correct, we had twenty days until the green maize was mature enough to eat, something we lovingly called *dowe*. It's the same as American "corn on the cob," when the kernels are soft and sweet and so heavenly in your mouth. Standing in those fields in February, I felt like one of the old explorers I'd read about—lost on the ocean and dying of thirst. Water everywhere and not a drop to drink. All day and night, I dreamed of *dowe*.

Toward the end of the month, Radio One said the *dowe* was ready in Mchinji, about 120 kilometers southwest. People started traveling there by the hundreds, including my uncle Ari, my mother's brother. Along the way he saw older men stop alongside the roadside and wave family members on, saying, "Go ahead. Get to Mchinji." He later heard those men had died. Gangs with spears and pangas guarded the fields and harassed those coming from other districts. Theft was rampant, and while my uncle waited outside Mchinji for his few kilos, he heard a great commotion on the outskirts of the village and rushed over to see. A mob had just attacked a thief and killed him. My uncle saw the man's body lying in the weeds, his neck cut to the bone.

After nearly five months of suffering, on February 27, the radio sent a message from our president. He was letting us know that the country was experiencing a hunger crisis. After consulting with his officials, he'd finally concluded it was an "emergency." As I said, our president was a funny guy.

AT THE BEGINNING OF March, the maize stalks were as high as my father's shoulders. At this stage, the flowers told you everything. Once the red and yellow silk began to dry and turn brown, you could start checking to see if the *dowe* was ready. I pinched the cob very hard to feel the grain.

If they crushed under your fingers, it was too early. But if the kernels were firm, you knew it was time.

Every day that week, Geoffrey and I left the tobacco fields to check the *dowe*, making sure to always use secret codes. God forbid my sisters followed and discovered that it was ready.

"Geoffrey," I said, "let's go burn the wasps."

"Yah, the wasps!"

Walking up and down the rows, we pointed out cobs that looked nearly done.

"Look at this one," I said. "In three days it will be in my mouth."

"Let's get firewood ready. Can we do this?"

"I think we can."

Then finally one day I spotted a cob that appeared ready, pressed my fingers against its grain. It was firm.

"It's ready," I said.

"Yes, it is," said Geoffrey, feeling another. "And so is this one."

"That means our long-awaited day has finally arrived."

"For sure! Let's proceed!"

I ran through the rows pulling the ripe *dowe,* gripping them lovingly in my hands. Soon I had fifteen ears spilling over in my arms. I peeled back the first layers of husks and tied them together, then draped them in a chain across my shoulders. Passing the tobacco shelter, I grabbed the armful of sticks we'd set aside for a fire. The sight of Geoffrey and me running into the compound with necklaces of *dowe* nearly started a riot.

"It's ready?" my sister Aisha asked, eyes wide and excited.

"It's ready."

"DOWE IS READY!"

I ran into the kitchen and quickly made a fire. White smoke soon filled the room and burned my eyes, filling them with tears. But I didn't care. I was too excited. My sisters now crowded into the tiny kitchen, fighting for space.

"Let me see!"

"No, I was here. You wait!"

"All of you go outside," I shouted. "There's enough *dowe* for all!"

I didn't wait for the fire to die into embers. I placed several cobs directly on the flames, flipping them until the peels were brown and blackened just right. I didn't even wait until the other side was finished cooking. I pulled the cob off the fire, so hot it scalded my fingers. I then peeled back the steaming husks and began to eat. The kernels were meaty and warm and filled with the essence of God. I chewed slowly and with great satisfaction, knowing I'd waited for so very long. Each time I swallowed was like returning something that was lost, some missing part of my being. When one side was eaten, I tossed the other back on the fire and moved to the next one.

My parents now gathered in the kitchen as I prepared the steaming food.

"I don't think this *dowe* is ready," said my father, snatching one quickly. "Let me have a taste." He ripped off the silk and bit into the *dowe*, savoring it as I had. Within seconds, the blood of life seemed to rush back to his face. He knew we would live.

"It's ready," he said.

That afternoon, Geoffrey and I probably ate thirty ears of maize.

As if heaven were coming down, the first pumpkins in our fields were also ready. For weeks I'd watched them closely, waiting for them to grow into the right shape and color. Now after all that time, here they were, as big as a man's head and orange as the morning sun. We cut them open and boiled them in a pot, seeds, skin, and all. My mother filled a basket with heaps of the warm, steaming meat and we devoured them right there. My God, to have a stomach filled with hot food was one of the greatest pleasures in life. Even Geoffrey came over and ate our pumpkins and *dowe*. Soon the swelling in his legs went away, and he began to smile again like his old self.

For me and Geoffrey, March was one big celebration. Each morning before working in the fields, we picked a handful of *dowe* and made a fire beneath the tobacco shelter, and ate a hearty breakfast.

"Mister Geoffrey, this is mine and this is yours."

"Sure, sure, give me mine."

I remembered a parable that Jesus told the disciples, the one about the sower of seeds. The seeds sown along the path get snatched away and damaged, those sown on rocky soil take no root and die, and those sown in the thorns get choked by the barbs. But the seeds sown on fertile soil will live and thrive.

"Mister Geoffrey, we're like the seeds planted on fertile soil, not on the roadside, stepped on by those walking past."

"No, no, we lived."

"That's right, Mister Geoffrey. We lived, we survived."

THE BASKETS OF *DOWE* and steaming pumpkins were like a great army marching to save us from certain defeat. In the trading center, people began to smile and talk about the future now. Life wouldn't really return to normal until after we harvested, and at home the usual blob of *nsima* still greeted us each night at supper, but at least it was the start of something good.

I started taking long walks to see how Malawi was doing, to see who'd survived and how they were getting along. With *dowe* now ready in the fields, people everywhere were drying it in their yards and making *chitibu*, which is a sweeter kind of *nsima*. People were regaining their strength, and smiling faces now greeted me along the roads. The same people who'd wandered weak and feeble now headed home carrying children on their backs and great bundles on their heads. But with the famine still fresh on my mind, I expected them to ask the same question I'd been hearing from every stranger for months: *Ganyu? I'm looking for ganyu . . .*

Instead, it was our normal happy greeting.

"*Muli bwanji!*" How are you?

"*Ndiri bwino, kaya inu?*" Fine, and you?

"*Ndiri bwino!*" Fine.

"*Zikomo!*" Thanks for asking.

"*Zikomo!*" No, thank you.

At the trading center, people now walked around shaking hands with their neighbors, as if they'd all returned from a long, hard journey.

"Good to see you, friend," they said. "You're still around."

"I'm around. How did you manage yourself?"

"God was on my side."

The blessing of *dowe* allowed us to return to our lives, but it also brought thieves. Many farmers who'd traveled to Wimbe from other districts weren't benefiting from the *dowe* and pumpkins because they didn't have their own fields. So instead they took it from others. The people living in Gilbert's blue gum grove waited until it rained at night, then stole the ripe *dowe*. After two weeks of this, most of their fields were gone. The same happened to us. Each morning we walked the road that bordered our field and found it littered with green leaves and *dowe* gnawed to the pith, as if a battalion had feasted all night.

Horrible stories of revenge soon began circulating in the trading center. One day, I heard some boys talking about the crimes.

"I heard some farmers in Kenji caught some men in their fields," one boy said. "You know what they did? Took pangas and chopped off their arms, asking, 'Long sleeves or short sleeves?'"

"*No!*"

"My cousin caught a young boy stealing his *dowe*," said Gilbert. "He put a metal pole in the fire until it became red hot, then told the boy to grab it. He did!"

All this talk about revenge made me wonder about our own fields. Later that night, I asked my father how we should punish those who stole from us.

"Should we kill them?" I asked. "Perhaps call the police?"

My father shook his head.

"We're not killing anyone," he said. "Even if I called the police, those men would only starve to death in jail. Everyone has the same hunger, son. We must learn to forgive."

CHAPTER NINE

OST STUDENTS AT KACHOKOLO Secondary and Wimbe Primary stopped going to school during the famine. After I dropped out, Gilbert continued to go to classes, and he told me that each day fewer and fewer classmates showed up. The teachers would call recess around 9:00 A.M., then disappear themselves into the fields and trading center to search for food. By February, there was no school at all.

But as the *dowe* and pumpkins became ready, and the district slowly reclaimed its energy, students began returning to school and classes resumed. But because my family still couldn't afford my school fees, I was forced to stay home doing nothing. Besides some weeding, there wasn't even much work to do in the fields until the maize harvest—still two months away.

I started wasting time in the trading center playing *bawo*. Someone also taught me a wonderful game called chess, which I started playing every day. But chess and *bawo* weren't enough to keep my mind occupied. I needed a better hobby, something to trick my brain into being happy. I missed school so terribly.

I remembered that the previous year a group called the Malawi Teacher Training Activity had opened a small library in Wimbe Primary School that was stocked with books donated by the American government. Perhaps reading could keep my brain from getting soft while being a dropout.

The library was in a small room near the main office. A woman was sitting behind a desk when I walked in. She smiled. "Come to borrow some books?" she said. This was Mrs. Edith Sikelo, a teacher at Wimbe who taught English and social studies and also operated the library. I nodded yes, then asked, "What are the rules of this place?" I'd never used such a facility.

Mrs. Sikelo took me behind a curtain to a smaller room, where three floor-to-ceiling shelves were filled with books. It smelled sweet and musty, like nothing I'd ever encountered. I took another deep breath. Mrs. Sikelo then explained the rules for borrowing books and showed me the collection. I'd expected to find nothing but primary readers and textbooks, boring things. But to my surprise, I saw American textbooks on English, history, and science; secondary texts from Zambia and Zimbabwe; and novels for leisurely reading.

I spent the day combing through the books while Mrs. Sikelo graded papers at her desk. Despite the variety of titles, I left that afternoon with books on geography, social studies, and basic spelling—the same textbooks my friends were studying in school. It was the end of the term, and my hope was to get caught up before classes started again.

At home I planted a thick blue gum pole deep in the ground near the mango tree out front, then made my own hammock out of knotted maize sacks. For the next three weeks, I began a rigorous course in independent study, visiting the library in the mornings, and spending the afternoons reading in the shade.

Gilbert offered to help right away. Each day I met him after school and he told me what they'd done in school.

"What did you cover today in geography?"

"Weather patterns."

"Can I get your notes tomorrow?"

"For sure."

"Thanks. I'll have the others back soon. I'm almost finished."

"No problem."

Reading on my own was often difficult. For one, my English was very poor, and sounding out words took a lot of time and energy. Plus, some of the material was confusing, and it would've helped to have a teacher to explain things.

"In agriculture," I asked Gilbert, "what do they mean by weathering?"

"It's when the rains disintegrate the rocks and soil."

"Oh, right, thanks."

ONE SATURDAY, GILBERT MET me in the library and we flipped through books we thought might be fun. I couldn't study all the time. One book that caught my attention was the *Malawi Junior Integrated Science* book, used by Form Four students. *Hmm,* I thought, and flipped it open. There were lots of pictures and diagrams, which I found easy to understand. I saw pictures of cancer and scabies and children stricken with kwashiorkor, like so many who'd wandered the country. One picture had a man in a shiny silver suit walking on the moon.

"What's happening here?" I asked Gilbert. "Why is he wearing this?"

"Looks cold," he said.

Turning the pages, I came across a picture of Nkula Falls on the Shire River, located in southern Malawi. I'd always heard about the hydro plants, but never knew how they worked. Simply by asking enough people in the trading center, and at home, I knew that water from the river flowed down the country until it reached the ESCOM plant, where it produced electricity. But how and why this worked, I had no clue.

The book described how the water turned a giant wheel at the plant, called a turbine, which in turn produced the electricity. Well, this seemed like the same concept as the bicycle dynamo. Behind my house there was the *dambo,* and sometimes if the rain was heavy, we'd see a small waterfall. Perhaps if I had a dynamo, I could put it under the waterfall and make it spin. Then I'd have electricity. The only problem was the *dambo* was just a soggy bog the rest of the year, and even in the rainy season, I'd have to run very long wires to my house to enjoy my music. That alone would cost a farmer's pay for an entire year.

Gilbert and I found another book called *Explaining Physics*. To my delight, it also used pictures and illustrations, mainly from England, to answer some of the questions I'd had for a long time—such as how engines burned petrol in order to move. Another illustration explained how brakes in a car worked. I'd always wondered if cars used strips of rubber to stop the wheels, like a bicycle. The book said otherwise.

"Vacuum brakes?" I said. "I need to borrow this book."

"Yah, for sure."

But *Explaining Physics* was much more difficult to read than *Integrated Science*. It was filled with long and complicated words and phrases. Over the next week, I struggled with the text but managed to figure out every few words and was able to grasp the context. For instance, I'd be interested in an illustration labeled "Figure 10," so I'd go through the text until I found where "Figure 10" was mentioned and study the sentences around it. For words that had no translation in Chichewa, I'd write them down and go to Mrs. Sikelo.

"Can you look up the word *electroscope* in the dictionary?"

"Okay," she said. "Any others?"

"*Kinetic energy* and *diode*."

"I think you're going above and beyond your fellow students. They're not studying this stuff."

"I know, but I want to know about this."

"Well, keep it up. Come back if you need more help."

AFTER TWO MORE WEEKS of reading this book, I found the most fascinating chapter—the discussion of magnets. I knew about magnets because they were used to make the speakers in radios. I'd busted off a few and taken them to school as toys, moving little slivers of metal around through a piece of paper. But as I read further, I discovered that some magnets—called electromagnets—are used to generate electricity, specifically in simple motors, like those found in a radio.

I also knew about the magnet's opposing sides. If you had two magnets, one side would always fight the other, refusing to stick together. However, flip one of the magnets over and it will snap to its fellow magnet. This book explained that all magnets have north and south poles. The north pole of the magnet will always attract the south pole, while two similar poles push away from each other. Because of its liquid iron core, the earth itself is a kind of large bar magnet, with magnetic north and south poles. Magnets, just like the earth, have natural magnetic fields that radiate between the poles. These lines are invisible, of course, but if you were to see them, they'd appear like the wings of a butterfly. One end of the bar magnet will always be pulled toward the magnetic north pole of the earth. This is how a compass works—inside there's a small bar magnet that finds north and keeps you from getting lost.

The book also showed how to make magnets out of everyday objects, such as nails, wire, and dry cells. When electricity from a power source—such as a battery—passes though a wire, it creates its own kind of natural magnetic field around the wire. This magnetic field can become even greater if that wire is coiled around a good conductor, like a nail.

The magnetic field will become even stronger depending on the number of times the wire is wrapped around the nail. Because of this, electromagnets have many uses. They can be used as giant magnets to pick up cars and heavy pieces of metal, while smaller electromagnets help power the simple motors found in many things around us—radios, appliances, alternators in cars.

In a simple electric motor, a coil of wire on a shaft is housed inside a large magnet. When the coil is connected to a battery and becomes magnetized, it creates a kind of fight with the larger magnet. The push and pull between these two magnetic fields causes the shaft to spin. Take, for instance, the fan that we use in hot weather. The blades spin round and round because there's a fight going on inside the fan.

The books said that if a power source can create this spinning motion—like in the fan—it can also cause the opposite to happen. Just as electricity causes a coil to become magnetic, a spinning coil that cuts a magnetic field will itself produce electricity. When a coil of wire rotates inside a magnetic field, it will produce a pulse of current. If a wire is connected to the coil, you can capture that pulse.

This is called electromagnetic induction, and it produces alternating current, or AC, since the direction of the current is changing. With electricity, you also have direct current, or DC, which is mostly found in batteries. It flows in one direction from one end of a battery to the other, through a bulb (or whatever load it's powering) to complete the circuit.

But most electricity found in your home is alternating current, produced by a giant spinning coil somewhere in a power station. The best example of an AC generator given by the book was a bicycle dynamo.

"The movement energy is provided by the rider," the science book said. *Of course,* I thought. This is how spinning motion generates power, both in the dynamo and in the hydro plant!

I can't tell you how exciting I thought this was. Even if the words sometimes confused me, the concepts that were illustrated in the drawings were clear and real in my mind. The various symbols—those for positive and negative, dry cells and switches in a circuit, and arrows indicating direction of current—made perfect sense and needed no explanation. And through them, I was able to grasp principles like magnetism and induction and the differences between AC and DC. It was as if my brain had long ago made a place for these symbols, and once I discovered them in these books, they snapped right into place.

I kept this book for a month and studied it daily, most often while ignoring my independent studies. It was like delicious food, and I seemed to want to share my knowledge with everyone I encountered. In the video show one day, the only television working was a black-and-white one, and some patrons were getting upset.

"Make it color," one said. "I want to see the colors!"

"But it's a black-and-white television," the owner answered.

"It looks like the other one, so make it color. It's all the same."

"Excuse me," I interrupted. "It's really not the same. A color television uses three electron tubes and a screen with fluorescence. It's right here in my book."

AFTER ABOUT A MONTH, the school term finally ended and Gilbert was free to hang out. One morning we went to the library to kill some time—we often stayed for hours, just sitting in chairs and reading—but today Mrs. Sikelo was in a rush.

"You boys spend hours in here taking my time," she said, "but today I have an appointment. Just find something quickly."

"Yes, madame."

The reason it took so long was that none of the books were arranged properly. The titles weren't shelved alphabetically, or by subject or author, which meant we had to scan every title to find something we liked. So that day while Gilbert and I looked for a good read, I remembered an English word I'd stumbled across in one of my books.

"Gilbert, what's the word *grapes* mean?"

"Hmm," he said, "never heard of it. Look it up in the dictionary."

The English-Chichewa dictionaries were actually kept on the bottom shelf, but I never really spent much time looking down there. Instead I asked Mrs. Sikelo. So I squatted down to grab one of the dictionaries, and when I did, I noticed a book I'd never seen, pushed into the shelf and slightly concealed. *What is this?* I thought. Pulling it out, I saw it was an American textbook called *Using Energy,* and this book has since changed my life.

The cover featured a long row of windmills—though at the time I had no idea what a *windmill* was. All I saw were tall white towers with three blades spinning like a giant fan. They looked like the pinwheel toys Geoffrey and I once made as kids when we were bored. We'd find old water bottles people threw away in the trading center, cut the plastic into blades like a fan, then put a nail through the center attached to a stick. When the wind blew, they would spin. That's it, just a stupid pinwheel.

But the fans on this book were not toys. They were giant beautiful machines that towered into the sky, so powerful that they made the photo itself appear to be in motion. I opened the book and began to read.

"Energy is all around you every day," it said. "Sometimes energy needs to be converted to another form before it is useful to us. How can we convert forms of energy? Read on and you'll see."

I read on.

"Imagine that hostile forces have invaded your town, and defeat seems certain. If you needed a hero to 'save the day,' it's unlikely you would go to the nearest university and drag a scientist to the battlefront. Yet, according to legend, it was not a general who saved the Greek city of Syracuse when the Roman fleet attacked it in 214 B.C."

It explained how Archimedes used his "Death Ray"—which was really a lot of mirrors—to reflect the sun onto the enemy ships until, one by one, they caught fire and sank. That was an example of how you can use the sun to produce energy.

Just like with the sun, windmills could also be used to generate power.

"People throughout Europe and the Middle East used windmills for pumping water and grinding grain," it said. "When many wind machines are grouped together in wind farms, they can generate as much electricity as a power plant."

Suddenly it all snapped together. The blades on these windmills were driven by the wind, much like our toys. In my mind I saw the dynamo, saw myself with the neighbor's bicycle those many nights ago, spinning the pedals so I could listen to the radio, thinking, *What can spin the pedals for me so I can dance?*

"The movement energy is provided by the rider," the book had said, explaining the dynamo. *Yes, of course,* I thought, *and the rider is the wind!*

The wind would spin the blades of the windmill, rotate the magnets in a dynamo, and create electricity. Attach a wire to the dynamo and you could power anything, especially a bulb. All I needed was a windmill, and then I could have lights. No more kerosene lamps that burned our eyes and sent us gasping for breath. With a windmill, I could stay awake at night reading instead of going to bed at seven with the rest of Malawi.

But most important, a windmill could also rotate a pump for water and irrigation. Having just come out of the hunger—and with famine still affecting many parts of the country—the idea of a water pump now seemed incredibly necessary. If we hooked it up to our shallow well at home, a water pump could allow us to harvest twice a year. While the rest of Malawi went hungry during December and January, we'd be hauling in our second crop of maize. It meant no more watering tobacco nursery beds in the *dambo,* which broke your back and wasted time. A windmill and pump could also provide my family with a year-round garden where my mother could grow things like tomatoes, Irish potatoes, cabbage, mustards, and soybeans, both to eat and sell in the market.

No more skipping breakfast; no more dropping out of school. With a windmill, we'd finally release ourselves from the troubles of darkness and hunger. In Malawi, the wind was one of the few consistent things given to us by God, blowing in the treetops day and night. A windmill meant more than just power, it was freedom.

Standing there looking at this book, I decided I would build my own windmill. I'd never built anything like it before, but I knew if windmills existed on the cover of that book, it meant another person had built them. After looking at it that way, I felt confident I could build one, too.

I COULD PICTURE THE windmill I wanted to build, but before I attempted something that big, I wanted to experiment with a small model first. From studying the pictures in the book, I knew that in terms of materials, I'd need blades, a shaft and rotor, plus some wires, and something like a dynamo to generate electricity from the movement of the blades.

Geoffrey and I had used a regular water bottle for those pinwheels we'd made as kids, but now I'd need something stronger. I'd seen Mayless and Rose playing cricket with an empty plastic jar of Bodycare perfumed jelly, so I went and grabbed it. The jar was shaped like a round tub of margarine with a screw-top lid. It was perfect. Leaving the lid intact, I sawed off the bottom of the jar with a bow saw, then cut the sides into four large strips, then fanned them out into blades.

I poked a hole through the center of the lid and nailed it to one of the bamboo poles my father was saving for tobacco shelters. I planted the pole in the ground behind the kitchen. But the wind hardly moved this contraption at all because the blades were too short. I needed extensions.

The floors in our bathhouses often fill with water so we install PVC pipes to serve as a drain. Several years earlier, a bathhouse behind my aunt Chrissy's place collapsed, and they simply built another one beside it. I knew there was a piece of pipe still buried under the bricks, and after twenty minutes of digging around, I managed to pull it free. I sawed off a long section of it, then cut it down the middle from top to bottom.

I stoked the fire in my mother's kitchen, getting it very hot, then held the pipe over the coals. Soon it began to warp and blacken, becoming soft and easy to bend, like wet banana leaves. Before the plastic could cool, I placed it on the ground and pressed it flat with a piece of iron sheet. Using the saw, I then carved four blades, each one maybe twenty centimeters long.

I didn't have a drill, so I had to make my own. First I heated a long nail in the fire, then drove it through a half a maize cob, creating a handle. I placed the nail back on the coals until it became red hot, then used it to bore holes into both sets of plastic blades. I then wired them together. I didn't have any pliers, so I used two bicycle spokes to bend and tighten the wires on the blades. That's when my mother came up behind me.

"What are you doing messing in my kitchen?" she said. "Get these toys out of here."

I tried to explain about windmills and my plan to generate power, but all she saw were some pieces of plastic stuck to a bamboo stick.

"Even children do more sensible things," she said. "Go help your father in the fields."

"I'm building something."

"Something what?"

"For the future."

"I'll tell *you* something about the future!"

It was pointless to explain. What I needed now was a bicycle dynamo or some kind of generator, and I had no idea where I was going to find such a thing.

For two days I tried to figure out how I could get a dynamo. I knew I could buy one, but where would I get the money? A trader named Daud had one for sale in his hardware shop in the trading center. I'd seen it there in the months before the famine—it hung on a shelf, shiny and silver and wrapped in plastic. It was beautiful. I went back this time, and sure enough, there it was. Daud stood between us, dressed in his round Muslim hat and long robe. I tried my charm.

"Fine day, Mister Daud."

"Yes, fine day."

"Your family?"

"Oh fine, thank you."

"How much for that dynamo behind you?"

"Five hundred."

"Yes, but you see, I don't have five hundred."

He laughed. "You know how it works. Go find the money and come back. It will still be here. And if not, I can order another."

I could get money by working *ganyu*. At the time, guys were making one hundred kwacha unloading trucks at CHiPiKU store, so I headed there. If I worked a whole week, I could have enough. I waited outside the store all morning and into the afternoon. The sun was dreadful, and I had no water. Finally the owner walked outside and saw me.

"Why are you standing here?" he said.

"Waiting for trucks."

"Trucks come every day," he said. "Except Monday."

It was Monday.

THAT NIGHT AT HOME, I hit upon another idea. The bicycle dynamo was the ideal generator for the bigger windmill I wanted to create, and at some point I'd have to work for the money to buy it. But for this experimental model, I could get by with using a much smaller generator, and I knew just where to find that.

I walked over to Geoffrey's house and found him in his room.

"*Eh, bambo,* do you remember where we put that International radio-cassette player?"

"Yah, it's here someplace. Why?"

"I want to extract its motor and use it to generate electricity."

"Electricity?"

"Yah, from a windmill."

Every time Gilbert and I had visited the library, Geoffrey was too busy working the fields. He hadn't been that interested in going anyway.

"We're headed to the library," we'd say. "Wanna go?"

"Go ahead," he'd answer. "Waste your time."

But now, when I told him my idea of building a windmill that would produce power—and then showed him what I'd built so far—he saw things differently.

"Cool! Where did you get such an idea?"

"The library."

USING A FLATHEAD SCREWDRIVER we'd hammered out of a bicycle spoke, I removed the screws from the International radio cover and tossed it aside. I removed the cassette deck, and behind it, I found the radio motor. It was half as long as my index finger and as round as a AA battery. A piece of metal stuck out from the top like a stem, attached to a small copper pulley wheel that hoisted a thin rubber belt.

I carefully extracted the motor. With some wire, I attached the motor to the bamboo so the copper wheel and the Bodycare lid were locked together side by side like two gears. But when I spun the lid, it slipped against the copper wheel. I needed some friction to make them catch.

"What we need is some rubber," I said.

"Where do we get it?"

"I don't know."

"What about the heel of a shoe?"

The rubber from our flip-flops was too spongy and not durable enough, otherwise we'd be set. Everyone wore those. We needed a different kind of rubber—the kind Geoffrey had mentioned that's used on the durable rubber flats worn by many women in Malawi. But it wasn't going to be easy to find those shoes. A company called Shore Rubber had been going through the villages collecting women's old rubber shoes to recycle and make new ones. They were offering a half kilo of salt for each pair, so of course, most women were taking the deal. But this was the perfect rubber for my windmill, and I was determined to find some.

All day Geoffrey and I dug through the rubbish pits at his house, Aunt Chrissy's house, Socrates' house, and finally my house, looking for shoes. Finally, after hours spent rummaging through mango skins, groundnut shells, banana peels, Geoffrey held up a shoe. One shoe.

"*Tonga!*"

The black shoe had been buried for so long it was now gray and covered in a layer of dirt and muck.

"Good job, chap!" I said.

Using my iron-sheet knife, I carved a tiny O-shaped piece out of the rubber, small enough to slide over the motor's copper wheel like a cap. This took me an hour. Once I'd pressed it on, the two gears caught just right and spun together.

The next step was to test the motor to see if it produced current. With Geoffrey spinning the blades by hand, I took the motor's two wires and touched them to my tongue.

"Feel anything?" asked Geoffrey.

"Yah, tickles," I said.

"Good."

Without stealing a radio from my father, the only functional radio we had was Geoffrey's Panasonic, which he used when he was working in the fields. Geoffrey loved his Billy Kaunda music, and often I'd sneak up on him in the maize rows and catch him dancing.

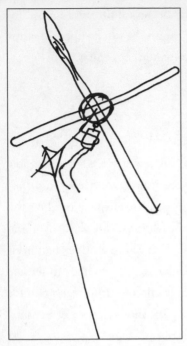

The first experimental radio windmill that I made with Geoffrey. The success of this model gave me the inspiration to go bigger.

I held the pole and fan in place while Geoffrey popped off the Panasonic's battery cover and removed the cells. Without them, we'd need to connect the two wires to the battery's positive and negative heads. Using my knowledge from the books, I assumed that because the radio operated on batteries, its motor produced DC power—unlike the dynamo on my father's friend's bicycle I'd played with months before, which produced AC power and only worked when connected to the radio's AC plug.

"How do I know which head is positive or negative?" Geoffrey asked.

"If you connect the wires and hear music, you got the right one."

"Whatever you say. Here it goes."

He pushed the wires inside, then twisted them so they connected to the heads.

"What do we do now?" he asked.

"Now we wait for the wind."

Then, just as I was saying that, the wind began to blow. My blades started to spin and the wheel began to turn. The radio began to pop and whistle, and suddenly there was *music*!

It was my new favorite group, the Black Missionaries, on Radio Two, singing their great song, "We are chosen . . . just like Moses . . ."

I jumped up so high I nearly disconnected the wires.

"*You hear that, man?*" I screamed. "We did it! It actually worked!"

"At last!" Geoffrey yelled out.

"Now I'll go even bigger. Superpower!"

"Yah!"

With that little success, I started planning for an even bigger windmill.

That model had already revealed itself in my mind, so there was no need to draw it out. The picture was of a much larger machine, which still used PVC pipe as blades. For the rotor, I'd need some kind of strong, flat metal disk. For the shaft, I'd use the bottom bracket, or axle, of a bicycle— what we called the "gav"—that connected the crankset and allowed the sprocket, cranks, and pedals to rotate. My plan was to cut the entire crankset off a bike to reduce the size, while leaving the back wheel intact. The blades would somehow be attached to the gav and act as pedals. When the wind blew, the blades, sprocket, and chain would spin, rotating the back wheel, and turning the dynamo.

I'm not sure where I got my confidence, but I knew this was a great plan. However, I had none of the materials I needed or the money to buy them. So to pull this off I had to do what I did with my radio experiment— I had to go and find them on my own.

For the next month, I woke up early and went in search of windmill pieces like exploring for treasure. The materials had to be strong and metal, and the best place to look was the tobacco plantation at Estate 24, just across from the Kachokolo school. The estate had an abandoned garage and scrapyard that was littered with machine parts and stripped bodies of cars and tractors, all forgotten and left to rust. I'd gone there a few times with Gilbert before the start of secondary school, after hearing the bullies at Kachokolo were even meaner than those at Wimbe Primary.

We'd looked for something to use for a weight bench in order to become muscular, like our great hero at the time—the Chinese martial artist Bolo Yeung from *Bloodsport,* a film the video show couldn't play enough for our satisfaction.

"Can you lift this metal like Bolo?" I'd say, straining against a rusty engine block.

"For sure," said Gilbert. "Stand aside."

We spent a few days trying to lift the heaviest pieces in the yard, but needless to say, the famine arrived and we never built the bench.

But now, I was returning to the yard to build my new machine. I took the long walk to Kachokolo, over the hills and streams. The land hadn't changed much since the start of the famine. The grasses were still high and starting to fade to brown, but the maize in the fields was now tall and green. Soon we'd be harvesting and our problems would be over, at least for this year.

I turned into the estate entrance once I reached the school. Within a few meters, I entered the scrapyard and stopped. Behold! Now that I had an actual purpose and a plan, I realized how much bounty lay before me. There were so many things: old water pumps, tractor rims half the size of my body, filters, hoses, pipes, and plows.

The stripped chassis of several touring cars lay bleached by the sun, in addition to two abandoned tractors. Their blue paint had long ago chipped and faded. The tractors had no tires or engines, just a rusted gearbox in their bellies. The steering wheels were still intact, as well as the gear shift and brake pedals. The clutch was a small button on the wheel next to a lever that worked as the gas pedal. The glass was busted off the instrument panels. The tall grass swallowed everything.

I've struck gold, I thought. Even better, the scrapyard was incredibly quiet and I was beautifully alone.

The first afternoon, I discovered a large tractor fan lying in the grass outside the garage. It was the ideal design for my rotor and would allow me to bolt the plastic windmill blades directly to the metal blades of the fan. That same day, I found a giant shock absorber, also from a tractor. I banged it against an engine block to knock off the casing. It fell away, revealing a piston I could easily weld to the fan to create the perfect shaft.

I needed some kind of ball bearing to connect the shock absorber and gav to reduce the friction. In order to find the right size bearing, I took a piece of rope and measured the end of the shock absorber, then went around to all the junk in the scrapyard and compared it to the various shafts with bearings attached. After three days I found the right match on an old groundnut-grinding machine.

I found a rusted coil spring and used it to bang the bearing loose. This took me all day because I had to take my time and be careful not to break it. After a couple of hours, the constant banging against metal caused me to get blisters on my hands that soon burst and bled. To endure the pain and keep banging, I imagined I was Bolo, burying his hands in hot sand to make them hard as steel. These mind games seemed to do the trick, and I was finally able to knock the bearing loose. My careful work paid off. It was in pristine condition.

As I mentioned, the scrapyard sat directly across from Kachokolo Secondary School, where I'd left a bit of my heart those many weeks ago. The school was empty now because the students were preparing for the second term, which started in only a few short weeks. I could see through the windows and imagined myself sitting in one of the empty desks.

Back at home, my father's tobacco was still drying under the shelters. Having made it through the famine, I hoped that once the crop was ready, it would fetch a good price at auction. We could finally settle our debts and pay my fees, and all my sorrow of being a dropout would melt away. I would return to school, and this time, I would be ready.

"Look out," I said. "Your man Kamkwamba will be back soon."

CHAPTER TEN

AS THE SCHOOL TERM approached, my father said nothing about me staying home due to lack of fees. In fact, one afternoon, he even handed me a couple kwacha to buy a brand-new Lion Brand Exercise Book and two pencils. My mother had also purchased a big bar of Maluwa soap, so a few days before school, I rolled out the half-tractor tire and scrubbed my uniform until the yellow stains faded in the suds. I took all this as a sign that things were normal once again. As you can imagine, with me dreaming about school every ten minutes, three weeks dragged past very slowly.

The night before classes started, I was so nervous I stayed awake for hours listening to the termites in the roof. It felt good to be waking up early for reasons other than farming. I'd desperately missed having the routine of dressing for school and waiting to meet my friends. But along with all these good feelings, I also had my worries: What if my independent studies hadn't been sufficient and my friends were too far ahead? Would they allow me to copy notes? Now that the famine was over, would the older boys be back and waiting for us? Who even survived the hunger, anyway?

When Gilbert appeared from the trees the next morning, I was incredibly happy to see him.

"Gilbert, bo?"

"Bo!"

"Sharp?"

"Sharp!"

"Fit?"

"Fit! Welcome back, friend, it's good to walk with you again."

"Oh thank you, Gilbert, it's good to be here!"

It was great to be back with my pals at school, all the jokesters and usual entertainers. I saw many of the same familiar faces. We were all still thin from the hunger, and that wouldn't change until the harvest, but at least our health was improving.

A few faces were missing, however.

"Where is Joseph, from Form Two?" I asked some people at recess. "Light complexion, short hair? I admired that guy."

"Oh, didn't you hear? He died."

A few others had died in the famine, but they were in other classes and I didn't know them.

As I had feared, I was far behind in everything: geography, agriculture, social studies, all the things I'd studied in the library. My classmates were studying graphs and variables and scientific names of animals. I didn't know any of that stuff. I struggled terribly for the first two weeks, copying all the notes I could, while also trying to get the hang of classes once again. It had been awhile, and so much had happened.

After about ten days, the grace period for fees was nearing an end and I started getting nervous. Something didn't seem right. My father knew my fees were due, but he hadn't mentioned anything. And fearing the worst, I was too afraid to bring it up. The closest we got was a brief conversation one afternoon in the fields:

"So how is school life?"

"It's going okay, but I'm so behind. I think with time I'll catch up."

"Well, just work hard."

That conversation had seemed normal, but I still couldn't help the sick feeling in my stomach each day I went to school. At the end of two weeks, we gathered in an empty classroom for morning assembly, and Mister W. M. Phiri addressed us, wearing his usual sweater and tie.

"Fees for this term are due Monday," he said, "and those students who didn't pay last term's fees must also pay those without delay."

It was like that. Even though I'd dropped last term, I still had to pay those fees if I wanted to continue. Together, the two terms equaled around two thousand kwacha. I actually hadn't realized this, and I was certain my father hadn't either. Given what we'd just been through, two thousand kwacha was an impossible amount. I knew my fate was sealed.

But instead of going home to ask my father for the money, for the next two weeks I tried to go to school for free.

I had to calculate my movements carefully. On Mondays and Fridays, W. M. Phiri held assembly inside the same classroom. There he read aloud all the names of students who'd already paid their fees, telling them, "Go to class straightaway." The students who remained in the assembly weren't allowed to enter the classrooms unless they produced a receipt. It was so embarrassing.

Geoffrey had been humiliated like this two years before, so I was ready. On the first day of roll call, I arrived at school with Gilbert as usual, but once the other students began filing into assembly, I ducked into the latrines at the edge of the school grounds. I stayed low and peeked out the tiny window. When everyone was released to join their classes, I slipped into the crowd like a cat in the henhouse.

Once in class, I sat in the back corner of the room with my head down. I was so scared of getting caught, I never asked questions for fear of looking suspicious. *As long as I'm silent,* I thought, *I can listen and still learn.* I was certain Mister Tembo was wise to my tricks, remembering that I was booted the previous term for lack of fees.

Several students got nabbed without receipts and were publicly expelled, making me incredibly nervous about this game I was playing. In the mornings I got awful stomachaches; it was so bad one day that I almost confessed to my father and ended it all. Gilbert would meet me on the road and we'd joke about my cunning tricks.

"Good morning, friend. I'm happy to see you're trying your luck again."

"Yah, let's hope today is not the end."

"Just stay quiet and keep your head down."

"I guess."

Finally, after two weeks, the teachers caught on to me. That morning, Mister Tembo read aloud the names of debtors in class, and that's when I was caught. The second my name was called, I stood up and walked to the door.

"Guys, I paid . . . just forgot my receipt," I said. "Don't worry, I'll get it and come right back . . ."

Once outside, I nearly started crying. I went home and told my father the news.

"I've been expecting this," he said. "I just didn't know when."

Instead of breaking my heart, my father went to see Mister Tembo and pleaded on my behalf. The tobacco would finally be dried and ready in a few weeks' time, and after paying his creditors who'd given us maize against our crop, my father was hoping against hope there'd be enough to sell at auction and pay for my school.

"I'll have the money soon," he pleaded. "Just please let him stay."

Mister Tembo spoke to a few other teachers. They agreed to let me to stay in school for three more weeks, long enough for my father to sell the tobacco.

These three weeks were fantastic, like winning the jackpot! No more sneaking around, no more butterflies in my stomach. Now I could relax and learn and participate in class. Now when the teacher cracked a joke, I laughed at the top of my voice.

"Oh, that's so funny!"

"Good point!" I'd say, whenever he made one. "I didn't know *that*!"

The other students gave me strange looks, but I didn't care.

"These past weeks he's playing the cool, quiet guy," they said. "But look at him now. What's happened?"

At the end of three weeks, the tobacco was finally dried and ready, turning a light chocolate brown in the sun. And once this happened, the cock came home to roost: the creditors began turning up at our house, looking to be paid.

"I've come for my fifty kilos," one said.

"Given our earlier agreement, do you have my twenty kilos?" asked another.

By the time the last trader left pushing a bicycle laden with our tobacco, only one sixty-five-kilogram bale was left hanging under the shelter. My father loaded it into a pickup and took it to Auction Holdings Limited in Lilongwe, where he received around eighty U.S. cents a kilo. But out of that sixty-five kilos of tobacco, only about fifty were a worthy grade for the floor. After transport costs and government taxes (about 7 percent), my father came home with around two thousand kwacha. It was just enough to cover my school fees, but then there'd be nothing left for the necessities of home, such as cooking oil, salt, soap, or medicine if someone got sick. Once again, we were broke.

My father tried negotiating again with Mister Tembo, but W. M. Phiri had already forbidden me to return. The Minister of Education was visiting various schools to ensure that all the students had paid their fees.

"If we're caught," said Mister Tembo, "some people could lose their jobs."

I was sitting on a chair in the yard when my father returned with this bad news. His eyes were pale and troubled, as if he'd wrestled with a ghost. I recognized the look on his face. It was one I knew well.

"I've done my best," my father said, "but the famine took everything."

He kneeled down to face me. "Please understand me, son. *Pepani, kwambiri.* Your father tried."

It was too difficult to look at him. "*Chabwino,*" I said. "I understand."

At least with daughters, like my sister Annie, a father can hope they'll marry a husband who can provide a home and food, even help them continue their schooling. But with a boy it's different. My education meant everything to my father. That night he told my mother he'd failed his only son. "Today," he said. "I'm a failure to my whole family."

I couldn't blame my father for the famine or our troubles. But for the next week I couldn't look him in the eyes. Whenever I did, I saw the rest of my life.

My greatest fear was coming true: I would end up just like him, another poor Malawian farmer laboring in the soil. Thin and dirty, with hands as rough as animal hides and feet that knew no shoes. I loved my father and respected him deeply, but I did not want to end up like him. If I did, my life would never be determined by me, but by rain and the price of fertilizer and seeds. I would do what every Malawian was supposed to do, what was written by God and the constitution: I would grow maize, and if I was lucky, maybe a little tobacco. And years when the crops were good and there was a little extra to sell, perhaps I could buy some medicine and a new pair of shoes. But most of the time, I knew, there would be hardly enough to simply survive. My future had been chosen, and thinking about it now scared me so much I wanted to be sick. But what could I do? Nothing, only accept.

I HAD NO TIME to wallow and grieve. The maize was ready, and my father needed all hands in the fields. I went into the harvest with a mixed heart. I was so convinced I'd never go to school again that entering the maize rows seemed like surrender, like walking into prison and locking my own door. But at the same time, my God, we were finally harvesting our food.

Harvests were always wonderful occasions, a time to think about all the mornings you awoke at 4:00 A.M. with the spiders in the toilet and hyenas in the fields—all the hard work of digging ridges, planting, and weeding and long days in the sun. We harvested all day with a satisfied mind, and at night we slept like a lion with a belly full of food. Harvests are a time to remember your sacrifice.

This time around, after two years of turmoil, of walking the desert in search of Canaan, it was as if God had led us out of bondage and revealed our great reward: we had a beautiful crop of maize, the best we'd seen in years.

For two weeks the work never stopped. First we walked down the rows with our pangas, chopping the tall stalks to the ground. Another person came behind, gathering together five to ten stalks and laying them across the rows. When this was finished, we collected the small heaps and piled them together into larger heaps called *mkukwes,* which stood upright and leaned together. This was to prevent the termites and mice from eating the cobs.

At the end of that month, we had four giant *mkukwes*—the most we'd had in many seasons. My father and I stood together and admired this beautiful bounty.

"It's unbelievable," I said.

"I know," said my father. "Even after eating and losing all that *dowe,* we're still blessed. Look at all this maize."

"What a harvest!"

We ripped the maize cobs off the stalks and stacked them into a pile, then hauled them to our house by oxcart. The man who owned the ox was paid in maize, as was the shopkeeper who then sold us the insecticide to keep off the weevils. For the next several weeks, we spent entire days sitting in the courtyard with a pile of cobs plucking off the grains and putting them into bags. We listened to the radio. We talked about weather. It was life brought back to normal.

In storage, our grain sacks were full once again, leaning fat and heavy against the wall, so many that they reached the ceiling and spilled out the doorway. Some soybeans from our garden had also matured, which meant we could now eat regular meals. Slowly, all the weight we'd lost during the famine started coming back.

"Ay, Papa," my mother said to my father, "you were looking *so skinny.*"

"And you, Mama," my father joked, "I see you're finally coming back to us. But William, *eh,* I was worried a strong wind was going to carry that boy from the fields."

We all laughed about it now, because it was only during better times that we truly acknowledged the bad ones.

WITH THE HARVEST FINALLY over, I was able to return to the scrapyard and continue searching for windmill pieces. I'd find one thing in the grass, pick it up, and think, *Now what is this?* only to spot something else that interested me even more. One day I was looking in some weeds and found the differential of a four-wheel drive. Using my screwdriver, I pried it open and discovered loads of fresh black engine grease. I scraped it into a plastic bag for future use. I also found cotter pins and tangled bits of wire, in addition to things I'd probably never use—brake pedals, gear levers, and the crankshaft of a small car engine. I took them all home anyway.

I was lucky because one of the biggest pieces I needed was right under my own roof from the start. My father had a broken bicycle and kept it against the wall in our living room, going no place. It had no handlebars, only one wheel, and a frame that was as rusted as anything in the scrapyard. I'd offered to repair that bike many times, but my father always said the same thing: "There's just no money."

The day I decided to ask my father if I could use his bicycle, I sat him down and explained the entire process, how the frame would make the perfect body of the windmill and would be sturdy enough to handle strong winds. The wind and blades would act as pedals and rotate the shaft and sprocket, and the chain would spin the tire and power the generator.

"Electricity!" I said, smiling big. "Water!"

My father just shook his head and said, "Son, please don't break my bike. I've already lost so many radios. Besides, one day we'll use that bike."

I thought, *Use it for what?* To ride seven kilometers to buy kerosene when you could have lights for free? Oh, it took *so long* to convince my father to give up that bicycle. I must have begged for an hour, explaining the process once more, almost putting together the radio windmill again just to remind him.

"I have a plan," I said. "Allow me to try. Just think, we could have lights! We could pump water and have an extra harvest. We'll never go hungry again."

He considered this a while, and finally gave in. "Okay," he said. "Perhaps you're right. But please don't mess it up."

I was so happy to have that bicycle. I took it to my room and leaned it against the wall near the other parts. With all the stuff I was collecting, it didn't take long for my room to appear like the scrapyard itself. All my cherished windmill pieces were arranged neatly on one side of the room—the shock absorber, tractor fan, bearings, with smaller pieces separated from large pieces for easy inventory. The rest of my room was covered in the extra junk I'd been gathering. Piles of random metal and old parts collected in the corners, around my bed, and behind my door. I never knew what I might need.

I forbade my sisters from sweeping or cleaning my room, because I was afraid they wouldn't appreciate my treasures and would brush away something important.

"But we need to scrub the floors," Aisha protested.

"Forget it," I shouted. "No one is allowed. I'll let you know the time!"

When I wasn't in the scrapyard digging for treasure, I'd hang out at the library or sit in my hammock and read. Even if my father didn't fully understand my windmill, he felt so bad about my schooling that he no longer forced me to work in the field. This made my sisters jealous.

"Why does William get to stay home and not us?" Doris asked my father one day. "Is it because he's a boy and we're girls? If he's staying home, so are we!"

"William has a project," my father said. "And if he's really wasting time, he'll be proven wrong eventually. You girls just worry about yourselves and go to work."

"Yes, Papa," they said and huffed away.

So with my father's blessing, the mornings and afternoons became my time to study. As I planned my windmill, I pored through chapters in *Explaining Physics* about electricity and how it moves and behaves and how it can be harnessed. I reviewed sections on home wiring, parallel circuits versus series circuits, and more stuff on AC and DC power. Going to the library, I renewed the same three books over and over until one day Mrs. Sikelo raised her eyebrow.

"William, are you still preparing for exams? What are you up to?"

"Nothing," I said. "Just building something. You'll see."

MORE AND MORE, GOING to the scrapyard began to replace school in my mind. It was an environment where I learned something each day. I'd see strange and foreign materials and try to imagine their use. One thing looked like an old compressor, or perhaps it was a land mine. I found real compressors and shook them to hear the pieces rattling inside; then I would try to open them and investigate. My imagination was constantly working. One day I pretended to be a great mechanic, crawling on my back under the old rusted cars and tractors with the tall grass clutching me in its arms. I shouted up to the customer.

"Start it up! Let's see how she sounds . . . push the gas, don't be shy! *Whoa, whoa, whoa!* That's too much!"

The engine didn't sound right, so I gave it to them straight: "Looks like you'll need an overhaul. I know, I know, it's expensive, but it's life."

I shouted to my other mechanics, who were slacking as usual.

"Phiri, today you're doing oil changes!"

"Yes, boss!"

Another walked over shaking his head. Problems again.

"Mister Kamkwamba, boss, we can't fix this car. We've tried everything, but it still makes a noise. How can you advise?"

"Start it up. Hmm . . . yes . . . hmm . . . just as I thought. Injector pump."

"Thank you, sir!"

"For sure."

I climbed atop the old rusted tractors, pressed the ignition button with my foot, and pretended to drive. "Out of my way—your man Kamkwamba must work!"

I dug the ridges in my field using my tractor, making up for all the days I'd swung a hoe in the hot sun. Oh, how I wished one of those tractors had actually fired up and moved. If it had, I would have taken the whole scrapyard home.

NO MATTER HOW ENCHANTED I became in the scrapyard, my wonderful moods never lasted long. The students across the street in the schoolyard could easily see me banging away on metal. When I wasn't being careful, they even heard me talking to myself. When I walked out carrying my windmill pieces, they would yell out, "Hey look, it's William, digging in the garbage again!"

At first I'd tried to explain the windmill, but they just laughed and said, "*Iwe,* you're wasting your time. This junk is good for nothing."

Even days when I tried sneaking past, someone would spot me through the open window and yell, "There goes the madman off to smoke his *chamba*!"

Chamba is marijuana.

Luckily I did have a few supporters and well-wishers. But Geoffrey had accepted an invitation from my uncle Musaiwale to work at the maize mill in Chipumba, and that meant that Gilbert was the only person who didn't laugh. Finally I decided that whenever someone shouted from the schoolyard, "William, what are you doing in the garbage?" I'd just smile and say, "Nothing, only playing."

These students immediately told their parents about the crazy boy in the scrapyard, and soon my mother was getting an earful in the trading center. Now when I came home with my pieces, she stared at me and shook her head. One day she came into my room quite worried.

"What's wrong with you?" she said. "Your friends don't behave this way. When you go to Gilbert's house, you don't see this kind of thing. I mean, look at this room! It looks like a madman's room. Only madmen collect garbage."

That night she told my father, "He's never going to find a wife like this, and even if he does, how will he care for her and feed his family?"

"Leave the boy alone," my father said. "Let's see what he has up his sleeve."

OVER THE NEXT FEW weeks, my scrap pieces kept revealing themselves like a magic puzzle. At one point, I realized I needed more PVC pipe, so without Gilbert's father looking, we dug out the drainage pipe from his shower stall. Inside, it was covered with several inches of slime that I had to scrape with my fingers. It smelled horrible.

Once it was clean and dry, I took the pipe home and cut it down the middle with a bow saw. Next I made a long grass fire behind the kitchen, then tossed the pipe atop the flames. When it began to bubble and curl, I rolled it off and pounded it flat. I then cut four blades at four feet each. I wanted to go ahead and connect them to the tractor fan rotor, but I had no nuts and bolts. So I spent two weeks in the scrapyard searching every piece of metal. But I had only one size wrench, which was too large for most of the nuts on the machines. To compensate, I wrapped a bicycle spoke inside the wrench hole and managed to loosen a few. However, most of them were so rusted they stripped against the tool or refused to even budge.

Gilbert then offered to help. He went to Daud's shop with fifty kwacha and bought a big bag of nuts and bolts. I was so grateful. But I still had no money to hire a welder to connect my pieces. Then one day while in the trading center, I had an amazing stroke of luck.

I was there playing *bawo* with some friends when a man pulled up in a truck. He was from Kasungu and needed some boys to help him load some wood.

"I'll pay two hundred kwacha for the job," he said.

I ran over waving my arms, "I'm ready, I'll do it," and the man told me to jump in back, along with about ten other boys. "Work hard, *ganyu* man!" the others yelled, knowing I was lucky. I spent all afternoon throwing logs into the truck and sweating in the sun, never so happy to be working in my whole life.

With my two hundred kwacha, I was able to pay a welder to connect the shock absorber to the sprocket to make it spin. I also needed him to melt holes in the blades of the tractor fan so I could attach the blades of my windmill.

Mister Godsten's shop was in the trading center, under a grass-covered shelter near the Iponga Barber Shop. Even though I had the money for the job, Godsten laughed when I walked up carrying my pieces.

"You want me to weld a broken shock absorber to a bicycle with one wheel?" he asked, mocking me. Several others were playing *bawo* under the fig tree and overheard.

"Ah, look, the madman has come with his garbage. We've heard about you."

"*Iwe,* he's not a man—just a lazy boy who plays with toys. He's *misala.*"

That meant crazy. I'd grown so tired of hearing these words.

"That's right," I said. "I'm lazy, *misala,* but I know what I'm doing, and soon all of you will see!"

They laughed at me anyway. I then turned to Godsten.

"To answer your question, mister," I said. "Weld the shock absorber to the bicycle. And make sure it's centered."

When the work was finished, I took the bicycle back to my room and leaned it against the wall. I could see why people were saying it looked like a madman's creation: the shock absorber jutted out from the sprocket like the arm of a strange robot, its joint fused with melted, blackened steel. My blades stood nearby, tall and beautiful, their white surface scorched and bubbled like the skin of a burned marshmallow. There were bags of bolts and nails and globs of grease hanging off the bike chain. The tractor fan looked like a dazzling orange star that would soon spin through the darkness. I couldn't wait to put them all together.

But even with this great design, I was still missing something, and it was a big something. Once again, I had all my pieces, but no generator. Where in the world was I going to find such an important and expensive thing? My family had no money, and I didn't dare ask my father to pay for the dynamo in Daud's shop.

I then thought of building my own simple AC generator. From all my research, I knew I'd need simple things such as a magnet, nails, and wire. But these materials weren't so easy to find. I didn't have the right gauge of insulated wire to use for my electromagnetic coils. I thought about busting apart a radio and taking the wire from its motor, but those motors didn't produce many volts, and as a result, the wire would be too short and thin.

I spent the next several weeks back in the scrapyard, turning over the rusted chassis of cars, the jagged slabs of sheet metal, and digging carefully through the tall grass, hoping I'd missed seeing a generator among the junk, perhaps an alternator I could take apart and use, or a bicycle dynamo. But no such luck. Unfortunately, I wasn't the only person looking for such materials. Some younger boys from the trading center had also discovered the importance of electric motors. But they just wanted to strip out the coils of copper wire and use them to sculpt toy trucks.

One day I caught a couple of them as I entered the scrapyard, and when I called out, "Hey you!" they took off running. I don't know why they were afraid. Maybe they'd heard stories about the *chamba*-smoking madman and feared for their lives. Anyway, when I reached where they'd been standing, I looked down and saw a perfectly good motor, stripped of its wires, lying there like one of those poached elephants without its tusks.

So with no generator, I began to fear my windmill would never be built. Every time I'd see a dynamo on someone's bike—usually broken or not even attached to bulbs—I'd think, *God, what a waste. Give it to me and I'll show you how to really use it!* I saw several dynamos during this period, but I didn't know these people so I never had the courage to stop them. Instead, I woke up each morning and looked at the pile of metal in the corner of my room, then went to help my father clear the fields. At night my windmill pieces were easier to look at, since everything disappeared in the dark.

ABOUT A MONTH PASSED. Then one Friday in July, Gilbert and I were walking home from the trading center.

"How's the windmill going?" he asked.

"I have everything, but still no generator," I said. "If I had this, I could build it tomorrow. I'm afraid this dream will never come true."

"Oh, sorry."

Just then, a guy passed us pushing his bike. I'd never seen this person, but he was around our age. As he went by, I looked down and noticed a familiar glitter by the tire.

"Look Gilbert, another dynamo."

However, by this point I'd stopped being bashful. I ran over to the guy and asked if I could see his bike. I bent down and gave the pedal a good spin, and when I did, the headlamp—an old car bulb—flickered on.

"It's perfect."

Gilbert turned to the guy. "How much to buy the dynamo?" he said.

"No, Gilbert," I said, "I don't have any—"

"How much?" Gilbert asked.

The guy refused at first, but finally gave in. No one was fool enough to refuse money at this time. "Two hundred kwacha," he said, "with the bulb."

"My father gave me some small money," said Gilbert. "Let's use it to buy the dynamo. Let's finish the windmill!"

Gilbert's father had given away all their food during the famine, and he wasn't farming as much because of his health. I was pretty sure their money was low. Still, Gilbert had bought my nuts and bolts for the rotor, and he now reached into his pocket and pulled out two hundred more kwacha—two red paper notes—and handed them to the man. After some messing around to get the dynamo and bulb off the bike, I was holding them in my hand.

"*Zikomo kwambiri,* Gilbert," I said. "Thank you very much. You're the greatest friend I ever had."

While Gilbert went home, I ran back to my room and placed the dynamo next to the other materials. It was like adding the last piece to the great puzzle in my life. The moment I did, a strong gust of wind blew open my door and spun a cyclone into the room, whipping up the pieces in its arms and revealing the finished machine, its blades spinning wildly through the blur of red dust. Or maybe that was only a dream.

CHAPTER ELEVEN

THE NEXT DAY AFTER lunch I began putting everything together. I took the fan, blades, bolts, and the dynamo outside behind our kitchen and arranged them in a neat row along the hard, barren dirt.

It was a wide, clear space to work and the perfect place to build my machine—close to both my room and the kitchen, which doubled as my laboratory, storage, and work sheds. It was also the best place for shade. When the midmorning sun was blazing, a big acacia tree behind the latrine cast a long enough shadow so I could tinker in comfort. Once the sun shifted in the afternoon, the kitchen provided good shade of its own. It was also the best place in the compound to receive the eastern winds that rushed over the mountains from the lake. As I began to work that afternoon, the Dowa Highlands were wrapped in blue sky and looked quite majestic.

The first thing I needed to do was connect the blades to the tractor fan, so I went to the kitchen and prepared my drill. I took the long nail with a maize-cob handle and stuck it into the embers of the fire. Once it was glowing red, I used it to bore four holes in the top of each plastic blade, then two more holes down the center. This process of heating, melting, and reheating took nearly three hours.

Taking a smaller bike wrench, I proceeded to fasten the blades to the

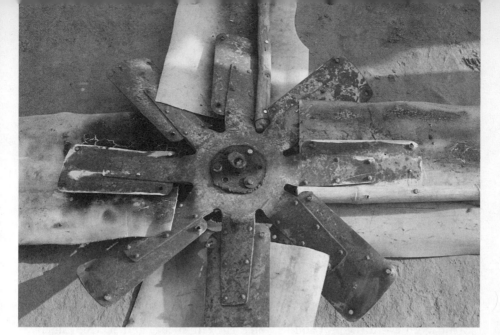

The tractor fan and blades from my first big windmill. This photo was taken after I'd replaced the Carlsberg caps with real washers and added some nails, but everything else is the same.

tractor fan with the nuts and bolts Gilbert had purchased. We didn't have proper washers to help secure the bolts, so I spend the next hour collecting bottle caps outside Ofesi Boozing Centre to use instead.

"Ah look," said one of the drunkards in the doorway, his corkscrewed eyes ready to tip his body over. "The government is finally cleaning the roads. Hey boy, how about a drink for an old man? I'm an orphan, you know."

"Sorry, I'm busy."

After I'd gathered sixteen Carlsberg caps, I brushed them off and hurried back home. I hammered them flat and drove bolts through their center. They worked perfectly. I then wired three-foot lengths of bamboo against each blade, like bones, for reinforcement. Once assembled, the wingspan of the blade system was more than eight feet across.

To attach the bicycle frame, I sat the rotor and blades atop four bricks, like putting a car up on blocks, so I would have room to work underneath. Now the challenge was to attach the bike to the blades. The bicycle was heavy and unbalanced, especially with the shock absorber jutting out from the sprocket. I managed to lift the bicycle up and turn it over to where the shock absorber pointed toward the ground, then rammed it into the center hole of the fan and blades. With the bicycle now balancing atop its shaft, I leaned under the blade system and slid the cotter pin through the other end of the shock absorber, locking it tight.

I fastened the dynamo onto the bike frame so its metal wheel hugged against the sidewall of the tire. I then restrung the chain over the front and back sprockets, making sure it was tight and would hold.

By now it was late afternoon and the compound was virtually empty. My sisters were off running errands, and my father was at a funeral in a nearby village. As I worked, I could hear my mother humming to herself in the kitchen as she prepared our supper, every once in a while whispering something to Tiyamike, who lay quietly on a pallet by the door. For once, the radio was turned off and I savored the rare silence and delicious smell of beans wafting from the kitchen. I had no interruptions.

By the time I finished attaching the dynamo and the chain, it was too dark to continue. I drew a bucket of water from the well and heated it for my bath, then went into the living room for supper. My sister Rose had returned from the shops and saw me on the way. She and my other sisters were standing around the machine and giggling.

"William, we haven't seen you all day," said Rose. "People in the trading center were asking about you."

"Your brother is a busy man," I said.

"I told them you were playing with your metals to make power."

"Something like that," I said, smiling. "Just wait. Soon you'll be in for a surprise, along with everyone else."

After eating my supper, I went to my room exhausted and fell fast asleep.

I WAS UP AT first light the next morning, ready to continue. I needed to figure out how to lift and move this heavy machine, so I took another long piece of bamboo and lashed it sideways through the frame with rope, making a kind of handle.

My plan was to build a tall wooden tower and put the machine on top of it, but first I wanted to see if it would actually work. For this I'd need a temporary tower. I had another piece of bamboo that was more than six inches wide. Using the hot-nail drill, I bored a hole through the top, then drove the pole into the red soil.

As I was doing this, I saw Geoffrey riding up on his bicycle. He was returning from Chipumba. It happened to be his day off and he was coming to visit me.

"*Eh* man, just in time," I said.

"Is this the same project you were working on?"

"Yah, this is everything. I'm glad you're here, friend. Help me lift this thing and fasten it to the pole."

We locked the wheel with a bent bicycle spoke to keep it from spinning, then carefully lifted the machine. Using wire and tire rubber, Geoffrey fastened it tight to the pole.

"Oh, look at it," he said.

I walked around the windmill, staring at it from every possible angle, as if it were a strange beast.

"It's beautiful," I said.

"Shall we?" he asked.

"We shall."

Geoffrey unhooked the spoke from the tire and the blades began to spin. It was slow at first, then got faster and faster. Within seconds, the blades began spinning so quickly the chain snapped in half and the pole nearly tipped over.

"Hold it!" I shouted, and Geoffrey and I caught the machine before it crashed to the ground and broke apart.

Luckily the hole I'd dug was a little too wide, and I was able to twist the pole and turn the machine out of the wind's direction. Once the blades stopped spinning, I began work on fixing the chain, a task that took two hours.

Part of this test run was to determine if the generator was producing enough current. To do this, I took the wires from the bicycle dynamo and jammed them into the AC socket of my father's four-battery International radio. The wind spun the blades around again, and for the slightest second, I heard music. It worked! But then the radio began pouring black smoke from its speakers and nearly caught on fire.

"Oh, no, your father's radio," Geoffrey said, looking around for my father.

I was too excited to care. "Did you see that power, Geoffrey?" I asked. "Did you see it?!"

The radio blew up because I'd forgotten the dynamo produced twelve volts, while the radio was only meant to handle half that power. Also, the wind spun the blades much faster than a person could pedal a bicycle, which caused a kind of power surge. I would have to figure out how to reduce the voltage.

I had read about voltage in *Explaining Physics,* and luckily, I had the book in my room (I was the only one who ever checked it out). I flipped back to where I'd seen a particular diagram of two separate bulbs being lit from a low, twelve-volt AC supply—exactly like mine. The bulbs were connected by long wires. One bulb glowed very bright, all because of something called a transformer, which boosted the AC voltage and made it stronger. But the second bulb didn't have a transformer, so energy was lost along its journey in the form of heat as it escaped through the wires. This was called dissipation.

Since energy is lost through wires when traveling long distances, I thought that perhaps I could use this idea with the dynamo and radio. I went back to my junk pile of radio parts and found an old motor, cracked it open, and took out the coil. I unwound the long copper wire and began wrapping it around a stick. I attached one end of this long wire to the dynamo and the other to the radio. After doing this, enough power was lost to play the radio without overpowering it.

FOR TWO DAYS THE windmill remained on the pole, hidden behind the house and out of sight. In the meantime, Geoffrey, Gilbert, and I set out to build my tower. In the early morning, we met in front of my house, grabbed an ax and pangas, and walked into the blue gum grove behind Geoffrey's house. This was the same forest where I'd been convinced I'd been bewitched by the bubblegum man, the same forest where I'd accepted magic and been defeated, and now I was back there to cut down trees to build a ladder to science and creation—something greater and more real than any magic in the land.

We walked slowly through the forest, looking carefully at each tree. Finally, we chose one that was about six meters high.

"Is this tall enough?" I asked, thinking aloud. "Strong enough?"

Gilbert and Geoffrey both nodded their heads.

"Then let's proceed."

The three of us tore into the trunk with our blades, and after ten quick minutes, the tree crashed to the ground. We then used the pangas to prune the branches and our hands to strip the bark. We felled two more trees just like the first one, and worked into the afternoon stripping them clean. By three o'clock, we were hoisting them onto our shoulders and carrying them home.

Just behind my bedroom, we dug three holes at one meter deep, each equidistant from the other. To fend off the termites, we wrapped the bottom of the poles in black plastic *jumbos* and buried them in the holes. Geoffrey had volunteered his pay from the maize mill to buy a bag of nails. Using the branches we'd cut, we began nailing reinforcements lengthwise like rungs of a ladder, starting four feet up so the children wouldn't be tempted to climb. Once the first rung was nailed in place, we'd climb and nail the next one, using the back of an ax as a hammer.

By sundown, the tower was built. It stood sixteen feet high and was steady, but from a short distance away, it appeared more like a wobbly giraffe who'd had too much *kachaso*.

"Get some sleep, gents," I said. "Tomorrow we raise the machine."

GILBERT AND GEOFFREY SHOWED up around seven the next morning.

The windmill's frame weighed about ninety pounds, and I knew the only way we would get it to the top was to use some kind of a rope and pulley. I didn't have a strong enough rope, so I used my mother's thick clothesline wire instead. Detaching the clothesline from its posts, we then fastened it to the windmill's bamboo handle.

Taking the other end, I climbed the tower and hooked the wire over the top rung, then dropped it down to Gilbert. Geoffrey stood below on the middle rungs to guide the machine as it went up. Standing there, I could see across the top of the acacia tree to where the patchwork of fields joined the highlands.

"Okay, Gilbert," I yelled. "Bring it up!"

Carefully, he began pulling the wire. First the windmill's handle lifted slowly, then the frame rose and wobbled into the air.

"Easy now!"

The three of us pulled at the wire with all our strength.

"Come on, guys," I shouted, "let me see your muscles!"

"I'm pulling all I can!" Gilbert said, straining with the clothesline.

"Don't let it slip, Geoffrey."

"You do your job and I'll do mine!" he replied.

Little by little, the windmill made its way up the tower. With each pull, it swung slightly and banged its cumbersome blades against the tower's wood frame. A couple of times it got stuck on the tower rungs, and Geoffrey had to knock it loose.

"Don't let the blades break!"

"I got it!"

It took about a half hour, but we finally had it close to the top. When the handle came within reach, I grabbed hold and screamed down to Gilbert: "Tie it down!"

Gilbert looped the wire around the base pole and the windmill held. Once I had a good grip, Geoffrey joined me at the top to secure it into place.

The day before, we'd drilled two holes in the wooden poles using hot bolts from a bike hub. (Drilling flat bolts into an eight-inch wooden pole with my cooking-fire drill takes a long time to do.) We'd also taken the bicycle to Godsten's shop and had him use his welding torch to blow two corresponding holes into the frame's crossbeam.

As Geoffrey pulled the bolts, washers, and nuts from his pocket, I held the windmill steady and aligned the holes. I could feel the machine straining in my hands.

"Make it fast, this thing is heavy!" I said.

"I'm trying. Just be tough and let me finish."

Geoffrey inserted the bolts and tightened them with the wrench. Once the windmill was fastened, we looked at each other and smiled. It felt sturdy and very strong. Sweat poured down my face and cooled with the breeze. I could hardly wait to watch the blades spin.

While Geoffrey made his way down the tower, I remained atop my perch taking in the scenery. To the north, I could see the iron-sheet roofs of the trading center and the brown row of huts that sat behind the main road. Then something strange began to happen. A line of people began trickling through the alleys from the shops and heading in my direction. They'd seen the tower from the market and were walking toward my house. Within a few minutes, a dozen people were gathered at the base. I recognized a few of the traders wearing their round hats and robes. One of them was named Kalino.

"What is this thing?" he asked.

Since there's no word in Chichewa for windmill, I used the phrase *magetsi a mphepo*.

"Electric wind," I answered.

"What does it do?"

"Generates electricity from the wind. I'll show you."

"That's impossible," Kalino said, smiling. Then he turned to get a reaction from the crowd. "It looks like a transmitter, and what kind of toy is that?"

"Stand back and watch."

I jumped down from the tower and ran to my room to get the final piece. That morning I'd found a thick reed and cut a section about ten inches long—the perfect size to hold the dynamo's small lightbulb. I then wrapped a long copper wire around the base of the bulb and strung it through the reed so that one end dangled out the side. This was my socket.

With the reed and bulb in my hand, I climbed the tower again and twisted its wires against the ones from the dynamo. As I did this, more and more people arrived. I watched the chorus below from the corner of my eye.

"What do you suppose he's doing now?" asked a farmer named Banda.

"This is the *misala* from the scrapyard my children spoke about," a fat man answered. "His poor mother!"

Looking out, I saw my parents and sisters hanging at the back of the crowd, eyes wide and waiting. Their jaws hung slightly open, as if there were seconds left on the clock and I had the ball. By now, my movements were automatic. I'd practiced this moment for months.

Aside from my family, about thirty adults had now gathered, and just as many children. They pointed at me.

"Let's see how crazy this boy really is."

"Quiet down! This is going to be a good show."

A steady wind whipped through the rungs of the tower, mixing the smells of chain grease and melted plastic. The bent bicycle spoke remained jammed into the wheel to hold it in place, but now the machine groaned against the breeze, as if begging me to release it.

Here it goes, I thought.

I grabbed the bicycle spoke and jerked it loose. When I did, the blades began to turn. The chain snapped tight against the sprocket, and the tire spun slowly, creaking and groaning at first. Everything was happening in slow motion. I needed it to go faster, immediately.

"Come on," I begged. "Don't embarrass me now."

Slowly, the blades picked up speed.

Come on, I thought, *come on.*

Just then a gust of wind slammed against my body, and the blades kicked up like mad. The tower rocked once, knocking me off balance. I wrapped my elbow around the wooden rung as the blades spun like furious propellers behind my head. I held the bulb before me, waiting for my miracle. It flickered once. Just a flash at first, then a surge of bright, magnificent light. My heart nearly burst.

"Look," someone said. "He's made light!"

"It's true what he said!"

A gang of school kids pushed through the crowd so they could see better.

"Look how it spins!" they said.

It was glorious light, and it was absolutely mine! I threw my hands in the air and screamed with joy. I began to laugh so hard I became dizzy. Dangling now by one arm with the bulb burning bright in my hand, I looked down at the eyes below—now wide in disbelief.

"Electric wind!" I shouted. "I told you I wasn't mad!"

One by one, the crowd began to cheer. They raised their hands in the air, clapping and shouting, "*Wachitabwina*! Well done!"

"You did it, William!"

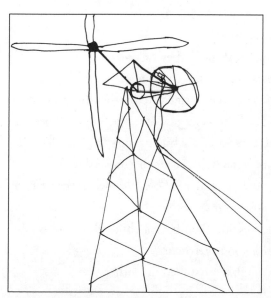

My first big windmill, measuring five meters (more than sixteen feet) and powered by the twelve-volt bicycle dynamo. My proudest creation.

"We doubted you, but look at you now!"

"I did it," I said. "And I'm going bigger now. Just wait and see!"

The adults began shouting questions up to me, but the noise of the blades ripping the wind behind me drowned out their voices. They crowded around Gilbert and Geoffrey instead, grilling them on the details. These guys couldn't stop smiling.

I stood up there for

about thirty minutes, taking in everything around me. It was a good place to stand and soak it all in. I only climbed down after the bulb became too hot with current and I had to let it go.

LATE IN THE AFTERNOON, I wired the bulb to the top rung of the windmill and left it. I was still high from the experience and needed to burn off some energy, so I went to the trading center to bask in the glory. From the market stalls, I could see the light at the bottom of the valley, flickering through the heat waves. I stood there for quite a while just watching it.

"What's that thing down there?" said a man nearby, clutching a sack of tomatoes. "It's catching the wind just like a helicopter."

The tomato seller was named Maggie, a friend of my mother's. "Well, the owner is this boy here," she said. "Why don't you ask him?"

"Is this true? How is that possible?"

I explained it to everyone the same way.

"I still don't understand," he said. "I need to come see myself."

For the next month, about thirty people showed up each day to stare at the light.

"How did you manage such a thing?" they asked.

"Hard work and lots of research," I'd say, trying not to sound too smug.

Many of these onlookers were businessmen from other districts who'd come to trade in the market. To the traveling traders, the windmill became a kind of roadside attraction, a must-stop while passing through Wimbe. Others came on bicycles from outer villages, with chickens and sacks of maize strapped to their bike racks with tire rubber. Women carrying flour on their heads stopped and spoke to my mother.

"God has blessed you," one said. "You have a child who can perform wonders. You'll never complain about kerosene."

The men who passed through approached my father.

"Your son made this?"

"Yes."

"What an intelligent boy. Where did he get such ideas?"

"He's been reading lots of books. Maybe from there?"

"They teach this in school?"

"He was forced to drop. He did this on his own."

That month I was busy clearing the fields and preparing them for the next growing season, working each day with joy in my heart. If I was in the field nearest the house, I'd pause in between swings of my hoe and just watch the blades spin.

When I came home after work each night, my mother told me, "Many more people came by today. They asked a lot of questions, but I can't explain. I told them to come back."

One night while playing *bawo* near the barbershop with Gilbert and Geoffrey, the trading center went black from another power cut. I quickly snuck home under the mask of darkness and connected the bulb, then ran back.

"Oh, I hate these power cuts!" a man said, leaving the barbershop and covering his head with a cap.

"Power cuts?" I asked with a smile. "What power cuts? Have you gents seen my house?"

Mister Iponga leaned out his shop, still clutching his dead clippers. "I think you take pride in our power cuts just so you can boast about your electric wind."

"Perhaps," I said.

AFTER A MONTH, I started working on running the lighting into my room. I needed a lot of electrical wire to do this, and as usual, I had no money to buy it. A couple of days later, Gilbert and I were hanging out at Charity's house near the trading center when I noticed meters and meters of insulated copper wire—just the kind I needed—being used as clothing line. In the corner of the room sat a large spool.

"*Eh,* man," I said to Charity. "How can you play around with wires like this when I need them so much?"

Charity said he'd been given the wire as payment for some work he'd done with his uncle, who was a trucker. He'd give me a discount because I was a relative. I told him I'd have to earn some money first, maybe find a job in the market or something.

"I'll go looking now," I said, but before I could get very far, Gilbert pulled out one hundred kwacha and gave it to Charity. Just like that, I had thirty meters of wire.

"This is just what I needed, Gilbert," I said. "Now I'll have lights in my room. Promise I'll pay you back."

"Don't bother," he said. "Just put lights in your room."

Once again, when all my hope was looking lost, Gilbert offered to help.

I RAN HOME FROM Charity's clutching the spool of wire. I flew down the trail, into the valley, then stopped near a clearing of trees just before my house. I could see the windmill, its blades spinning furiously. My stomach did a flip every time I saw it. I then took a deep breath and continued home.

I unspooled the wire from its wooden base and measured the distance between the dynamo and my room. Pulling out a few extra meters to be safe, I clipped it with my knife. With the wire in one hand, I climbed the tower. I unhooked the reed and bulb from the top rung, then pulled at the dynamo wires attached. I didn't want a shock, so I avoided touching both wires together. The wind was kicking up, spinning the blades so fast and close I was nervous they were about to give me a haircut. Finally, after some serious twisting and pulling, the wires snapped apart. I put the bulb and reed into my pocket, then twisted the new copper wiring to the dynamo wires, wrapping a black *jumbo* at the point of connection. I then climbed down.

The roof of my bedroom was constructed from several blue gum poles supporting a sheet of black plastic and layers of grass thatching. Leaning a ladder against the outside wall of my room, I located the middle beam and wrapped it several times with wire, leaving about two meters to spare. I then hooked the end of the wire to a long bamboo pole and pushed it along the roof beam, under the plastic and thatch, and into my room.

Once inside, I grabbed hold of the wire and, while standing on the

Me connecting the light bulb in my room. As you can see, it's just a small car bulb that dangled from my ceiling.

bed, wrapped it around the beam above. When I reattached the reed and bulb, the light powered on. I ran to the door and slammed it closed, then marveled at the new electricity. My window was only a sliver in the wall, giving no real light. With the door closed, the light illuminated everything. I saw the piles of metal scraps in the corners, the random nuts and bolts and wire clippings strung along the mud floor. For once, I had my own private, lighted space.

My sisters had seen me run inside, and now they were knocking and asking what was happening.

"Can we see it?" asked Doris.

"Come on in," I said.

That evening, I lay in bed and stared up at the bulb. It flickered yellow with the creaking sound of the blades outside, bright enough to see my hands and feet and the library books at my side. My parents and sisters all stopped by to see for themselves, squeezing into my tiny room to marvel at this new addition to our home.

"Look at William, staying up past the dark!" my father said.

"Congratulations, son," said my mother. "You know, we'd also like our rooms to have these lights. Think you could manage?"

"Oh," I joked, "are you sure you want to use electricity generated by a madman?"

"*Eh,* you proved everyone wrong," she said, smiling. "But I admit, I did worry about you."

"What if the wind stops blowing?" asked Rose.

"Well," I said, "the light goes off, and I'm stranded. But I'm already thinking of a plan for having a battery."

Once I had more wire and a car battery, I explained, I could store electricity for the times when the wind stops blowing. It could also provide light for the entire house. It would have to be done little by little, but once complete, it would save my parents the money they normally spent on kerosene, and that was just the beginning. The next machine would pump water for our fields. One day, windmills would be our shield against hunger.

That night, I was too excited to sleep. After everyone went to bed, I stayed awake and flipped through *Explaining Physics,* preparing for the next step. Every now and then, I stopped what I was doing and looked up at the light and the way it blinked and flickered. The warm glow painted the walls and the pages of the book and shined against the red clouds of dust pushed in from outside.

As usual, that night a strong wind was blowing.

CHAPTER TWELVE

*A*s I explained to Rose, the windmill wouldn't work without wind. On calm, quiet nights, I'd still be stuck in the dark looking for matches. The only way I could change this was to find a battery. But while I waited for one, I still managed to use my windmill for other purposes.

My cousin Ruth, who was Uncle Socrates' daughter, happened to be visiting from Mzuzu town. She was older and married and had a mobile phone, and often, she would bug me to take it to the trading center and charge it for her. A few guys in the trading center were making loads of money charging phones for people who didn't have electricity in their homes. These guys would cut deals with shopkeepers, who allowed them to string an extension cord out the door, where they kept a small shaded stall. They sold scratch cards for phone units, and a few even had mobile phones on a table and you could pay to make calls, kind of like an open-air phone booth. These stalls can be found all over Malawi. In the bigger towns like Lilongwe, some guys even powered photocopiers and electric typewriters this way, and let people prepare their résumés and then call about an available job—all on the sidewalk or the dirt road. Of course, the constant blackouts weren't so good for their businesses.

One day I was complaining about taking Ruth's phone to the trading center again when she said, "Why can't you just charge phones with your windmill? It produces electricity, doesn't it?"

I'd already thought about this, but I knew I didn't get enough voltage from my bicycle dynamo to power a charger. My dynamo produced twelve volts, but a charger needs two hundred twenty. Twelve volts was fine for a lightbulb, but not strong enough for heavier jobs.

I had already discovered that energy decreases as it passes through wire over long distances. I used that concept when testing the radio against the dynamo. But for this, I needed to build something called a step-up transformer.

Power companies across the world, especially in Europe and America, step up power all the time. Huge alternators generate electricity in a power station, but as my previous experiment demonstrated, it loses strength on the way to your home. To remedy this, the power company installs step-up transformers in places along the route that boost the power before sending it on its way.

A step-up transformer has two coils—the primary and secondary—that are wrapped side by side around a core. Alternating current flows back and forth and is constantly changing, and this change causes the primary coil to induce a charge in the secondary coil. This process is called mutual induction, which means that voltage from one coil jumps into another. The result is that the overall voltage is increased. I had learned about this in a chapter in *Explaining Physics* entitled "Mutual Induction and Trans- formers," which also featured a picture of a man with white hair and a bow tie. This was Michael Faraday, who invented the first transformer in 1831. What a good feeling that must have been.

Using the diagrams from this chapter, I was determined to make my own step-up transformer. First, I used some sharp pliers and cut an iron sheet into an "E." The diagram demonstrated twenty-four volts being transformed to two hundred forty. I knew that voltage increased with each turn of the wire. The diagram showed the primary coil to have two hun- dred turns, while the secondary had two thousand. A bunch of mathemati- cal equations were below the diagram—I assumed they explained how I could make my own conversions—but instead, I just started wrapping like mad and hoped it would work.

I connected the dynamo wires to the primary coil, while the secondary coil was wired directly to the prongs of a phone charger. I got quite a shock while I was tying those bare wires to the charger, but once they were attached, I was ready to plug in the phone. Ruth was standing over me, watching carefully.

"Don't blow it up," she said.

"I know what I'm doing," I said, lying.

I plugged in the phone jack, but instead of blowing up, the screen lit up and the bars began moving up and down the side. It worked!

"See," I said. "Told you."

To make things easier, I went a step further and made an electrical socket in my wall, just like at Gilbert's house. I had a radio that operated on both batteries and AC electricity. In these kinds of radios, one end of the power cord usually plugs into the AC socket, and the other end goes into the wall of your house.

To get electricity in my room, I simply transferred this system into my wall. I pulled out the entire AC socket from the radio and fixed it into a wall mount made from flattened PVC pipe. I cut a hole in my wall and installed the mount so it looked a like a normal electrical socket, then connected the windmill wires from the outside. I reattached the radio's power cord to the AC socket, then cut off the electrical plug that normally shoves into the wall. Taking those severed wires, I attached them directly to the phone charger. It was kind of backwards, but it worked. When news of the invention reached the trading center, the line of people arriving to charge their phones reached the road.

Most of the time people still pretended to be skeptical that I could do this. I think they were hoping I wouldn't charge them money.

"Are you sure this thing can charge my phone?"

"I'm positive."

"Prove it."

"See, it charges."

"My God, you're right. But leave it for a little while longer. I'm still not convinced."

After two months of using this method, I finally went bigger. I was always on the lookout for a car battery, then one day I noticed that Charity had one at his house.

"I found it on the road," he said. "It must've fallen off a truck."

Wherever it came from, I started begging and pleading with him, until finally he agreed to sell it to me on installments.

From studying my books, I knew that in order to charge the battery—which uses DC power—I had to first convert the AC power produced by the bicycle dynamo. My book talked about certain things called diodes, or rectifiers, which are found in many radios and electronic devices and convert the power for you. The type of rectifier I needed looked like a tiny D-cell battery on a long metal skewer and reminded me of the smoked mice that young boys sell on the roadside as snacks. Studying the picture, I opened one of my broken six-volt radios and easily found a diode. Once again, I fashioned a soldering iron out of a thick metal cable I heated in the fire, then fused the diode to the wire between the windmill and battery. It was so simple. I then replaced the standard AC phone charger with a DC model that plugs into a car's cigarette lighter—a gift from my cousin Ruth.

With the battery, I was also now able to install three additional bulbs in the house, all running on a parallel circuit. For some reason, the light from the bicycle dynamo still worked using DC power. However, regular incandescent bulbs used in most homes will only work with AC power, so I had to look for alternatives. I managed to find small car bulbs in Daud's shop that worked on DC—one was a brake light, and the others were headlamps. I installed one bulb outside, just above my door, another in my parents' bedroom, and one in the living room. When the battery was fully charged, it could store enough power for three full days, even when the wind was still.

I constructed a simple light switch for each bulb using bicycle spokes and strips of iron. For the toggles, I wanted a good nonconductive material that I could easily shape the way I wanted. So taking my knife, I carved out several round buttons from my old broken pair of flip-flops. I mounted these switches inside small boxes I'd made from melted PVC pipe.

A standard light switch has three separate wires: one from the power supply to the switch, one from the switch to the bulb, and one from the bulb to the power supply. Whenever I pushed the flip-flop button, the spoke and iron connected the terminals, completing the circuit.

Finally, I could touch the wall and get lights!

Not long after I'd completed this wiring, I walked into the living room one night and found my family sitting around, listening to the radio. My mother sat on the floor in the corner, crocheting a beautiful orange tablecloth. My father and sisters just stared ahead, lost in the news program on Radio One. I pretended to be one of reporters on the radio, barging in with my microphone.

"I'm now standing in the living room of the Honorable Papa Kamkwamba," I said, in a deep, serious voice. "Mister Kamkwamba, this room used to be so dark and sad at this hour. Now look at you, enjoying electricity like a city person!"

"Oh," said my father, smiling. "Enjoying it more than a city person."

"You mean because there's no blackouts and you owe ESCOM nothing?"

"Well, yes," said my father. "But also, because my own son made it."

HAVING LIGHTS AT NIGHT was a remarkable improvement in my family's life. However, it still wasn't perfect. The battery and wires I used were not the best quality. The wire Charity had given me had run out, so I was mainly using bits and pieces of wire I'd found in the scrapyard. Some of these wires were never meant for electricity, but I used them anyway. My patchwork of fused-together wires weren't covered in plastic insulation, but were bare and often sparked when I connected them to the battery terminals. Since I didn't have a proper conduit to house them, the positive and negative wires leading from the bulbs and switches were simply nailed to the walls and across the ceiling like a heap of Christmas lights. I took great care not to cross them, since my poor roof was made from wood and grass and could easily catch fire.

Worse, the blue gum poles that supported the thatching had long been infested by termites. Each night, I went to bed hearing the sound of their teeth nibbling away, only to wake up in the morning to little piles of sawdust on the floor. Their constant appetite had finally hollowed the beams, causing them to sag a bit. Having a mess of bare electrical wires strung across these beams made it even more dangerous. It wasn't long before it nearly caused an accident.

One afternoon after a big storm, I returned from Geoffrey's house and discovered that the beam had finally broken, probably from the wind pushing on it. My ceiling now sagged in the middle and my floor was covered in dirt and grass. The broken beam had also dumped hundreds of squirming termites onto the floor and across my bed.

At first I tried sweeping them off, but there were so many. By that time, my father had managed to purchase a few more chickens, and as I looked out my open door, I saw a gang of them walking past.

"Come here, chickens," I called out. "Do I have a treat for you!"

I tossed a few termites out the door to lure them in. Once they realized what a great bounty was waiting for them inside my room, they went mad with hunger. Soon my floor and bed were filled with chickens, squawking and flapping their wings with excitement as they pecked at the helpless termites.

"Get them all!" I shouted. "I don't want to see another one alive!"

The termite incident caused such a commotion that I didn't even notice the smell of something burned. After the chickens cleared out, I looked closely at the broken beam that hung from the ceiling and saw that my wires had crossed during the collapse. Instead of starting a fire, they were so cheap and thin they'd simply melted and snapped in two. I thanked God no one had been hurt.

Geoffrey came over and we stared at this mess together.

"*Eh, bambo*, it's a good thing I'm too poor to buy proper wire," I said. "If I'd used anything better, I'd have burned up my home."

"I warned you about the roof."

"Sure, sure, but I didn't listen."

I NEEDED A PROPER wiring system, and as always, I returned to my *Explaining Physics* book and found just the model. Page 271 had a diagram of a mains system from a home in Britain. The drawing showed that when the wires ran from the main power supply, the first place they led was to a circuit breaker, which worked to kill the power when the circuit became overloaded. I needed one of these.

The circuit breaker in the diagram used fuses, which contained tiny metal filaments that melted when the overload occurred. I didn't have any fuses, so like with everything else, I improvised by using what I had around my house. Instead of a filament and fuse system, I modeled my circuit breaker after the electric bell, which I'd also read about in the book.

An electric bell works when a coil suddenly becomes magnetized and pulls a metal hammer that strikes the bell. However, during this motion, the hammer also trips a switch and breaks the circuit, returning it to its original position. Of course, this happens about a dozen times per second.

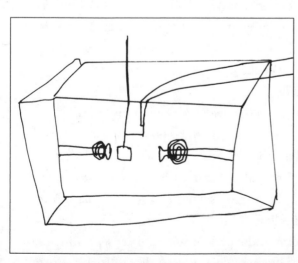

A drawing of my handmade circuit breaker, which I modeled after the electric bell.

I started my circuit breaker by constructing a breaker box using melted PVC pipe (as I'd done with my switches). I wrapped the heads of two nails with copper wire to create two electromagnetic coils, then mounted them inside the box, with the coils facing each other about five inches apart. Coming up through the middle, I placed a bar magnet (from a radio speaker) attached to a bicycle spoke that could easily flip from side to side. I then took a spring from a ballpoint pen and stretched it out. I placed it between the bar magnet and nail to where it rested lightly against the circuit's live wire. Basically, this spring completed the circuit and acted as a kind of trap.

When the light was turned on, the electricity flowed from the battery into this circuit and magnetized my two nails—one of which was situated slightly closer to my bar magnet. The polarity is determined by which direction the current runs in, and I'd wrapped the nails with wire so the one closest to the magnet pushed, while the other nail pulled. This pushing and pulling from both sides kept the bar magnet delicately balanced, never knowing what to do.

In the event of a power surge, this balance would be broken. The nail closest to the bar magnet would receive the surge first, pushing it vigorously into the other nail, knocking the pen spring loose from the wire, and breaking the circuit and flow of current.

Once this was built, I nailed the circuit breaker box to my wall just above the battery. Every night I sat on my bed and stared at the box, waiting for it to work. I got my wish about two weeks later when a cyclone hit my house.

I'd been in the trading center all day, and when I returned, I noticed something odd. Bits and pieces of my thatch roof were lying in the yard, and my roof itself looked as if a giant had given my house a good shaking. My mother came out from the kitchen.

"What happened?" I said.

"A big cyclone just came from the fields. We had to run inside."

When I went inside my room, I saw that my poor roof had collapsed again and bits of ceiling were all over the floor. I also noticed that the circuit breaker had flipped. I moved the magnet back to the middle, but it refused. I tried again, but the magnet swung back into the nail. After disconnecting the battery, I followed my wires along the ceiling and discovered they'd become tangled in the cyclone winds. After separating the wires and reconnecting the battery, the bar magnet flipped back between the two nails. Once again, I'd narrowly escaped a fire.

But I was more excited about my circuit breaker than anything.

"See, man, the house would be ashes right now," I told Geoffrey. "All of my clothes and blankets would be gone. The circuit breaker worked!"

"The circuit breaker is great," he said. "But I think a better solution would be to fix your roof."

ANY NEW INVENTION IS going to have its share of problems, and aside from the amateur wiring, one of my other headaches was the bike chain. Whenever the wind blew very hard and spun the blades, the chain would snap or simply jump off the teeth of the sprocket, forcing me to climb up the tower to fix it. This required stopping the blades, a task I hated.

One morning I was sleeping really well, one of those peaceful sleeps just before the cock crows. I think I was even smiling in my dreams when a terrible racket forced me awake. I knew just what it was. The chain had slipped again. I heard the wind whipping the acacia tree outside and my tower was rocking back and forth. The blades were spinning so fast they were buzzing on the rotor. If I didn't fix them, they'd eventually snap in two, flying through the air like daggers. This frightening thought was enough to push me out of my warm bed.

Earlier, I'd come up with a wonderful plan of installing a brake system that would stop the blades by simply pulling a handle. It would be similar to the coaster brakes that stop a bicycle by pedaling backwards. These brakes work by using a special hub, which contains a gear system that locks when backward motion is applied. The plan was to attach this hub to the shaft, then trigger it by jerking a cable from below—ideally one that ran into my room so I wouldn't have to get out of bed to operate it. But after weeks of searching, I couldn't find the right hub, and so I had no braking system in place. This morning I had to do it the hard way.

I climbed to the top of the tower and, as usual, kicked off my flip-flops so I could get a better grip. But the wind was violent and angry, pushing the tower from side to side so much that I thought it would tip. I wrapped my legs through the rungs and held on for life. But in trying to keep my balance, I didn't notice the bicycle frame swinging forward along with the tower. The next gust of wind sent the blades straight into my hand and knocked me off balance. I slipped and nearly fell, grabbing hold of the rungs and cursing. Looking at my hand, I saw the blades had shaved the meat off three of my knuckles, which were now dripping blood that scattered with the wind.

"You are my own creation!" I shouted to my windmill. "So why are you trying to destroy me? Please, let me help you."

Once I regained my balance, I pulled a thick strip of bicycle tire from my pocket that I'd cut for such repairs. Gripping it in the palm of my hand, I held my breath and grabbed hold of the spinning sprocket. I felt the jagged teeth cutting into the rubber like a saw blade.

"Stop!"

Once I managed to stop the blades, I shoved the bent bicycle spoke into the wheel to keep the machine from spinning, then reattached the chain. The next time this happened, I wasn't so lucky. The teeth on the sprocket finally cut through the bike tire and ripped my flesh. Then it happened again. I still have scars from this whole painful period.

DURING THIS TIME, GEOFFREY was still working with Uncle Musaiwale at the maize mill in Chipumba. He'd been hired to sweep the floor and fetch things for our uncle whenever he needed them. But once there, our uncle mostly drank, and Geoffrey was forced to run the mill on his own. He'd wake up early and do maintenance work on the machines, making sure they were filled with diesel and oil. Then he would open for business. At night, he shared a bedroom with Musaiwale, who'd come home from the bars singing merrily before collapsing into a deep, snoring sleep.

Chipumba was about twenty-five kilometers away, and almost once a month, Geoffrey would bike home and tell me about his hard life as a workingman.

A close-up of my windmill with the improved pulley system. As you can see, I kept the bike chain as a rope to remind myself of the pain and scars it caused me.

"They force me to ride up five hills to get diesel," he said. "And on the way back the fuel soaks my clothes. I'm telling you, brother, I'm missing you guys terribly."

But he also described how the grinding machines in the mill operated by using pulleys and rubber belts.

"You can get rid of this chain problem if you use a belt," he said. "They use them on the mill, and they never fail."

This was great news. A pulley was just what I needed to increase the tension between the front and back sprockets, which was the reason the chain kept flying off. And unlike a chain, a belt didn't have troublesome cogs that constantly required grease, which I'd long run out of.

Going to the scrapyard, I was easily able to find two pulleys from an old water-pumping engine. I used a piece of heavy steel and hammered at them for several hours, snapping their cotter pins and sliding them off the machine. But the center hole of the larger pulley was too big for my shock-absorber shaft to fit. I had to weld it alongside the sprocket itself.

These days, Mister Godsten the welder no longer made fun of me. Whenever he saw me walking up holding these random pieces, he just smiled and fired up his torch.

"Tell me where," he said.

Mister Godsten even let me use his grinder to flatten all those sharp teeth on the sprocket until its edges were smooth. It felt like sweet revenge watching them reduce to nothing under a shower of sparks.

"This is for all my scars!" I shouted.

Once I installed the pulleys that afternoon, they seemed to work great. The only problem was that I didn't have a proper belt. Looking around for something to hold me over, I cut the handle off an old nylon bag and rigged it around the pulleys. It worked for about ten seconds before slipping off. I then cut open a few batteries and removed the black jelly inside—a kind of tar that houses the cell's carbon rod—hoping that would work as an adhesive. But the tar simply wore away after a few hours. An old man in the trading center then gave me an actual belt from a milling machine— he was using it to fasten vegetables to his bike—but it was broken in half, forcing me to mend it constantly with my crochet needle and fiber from a car tire. That never lasted long. But with nothing else, I was forced to use this for two months. Even after all that trouble, I was still climbing the tower twice a day.

Finally, Geoffrey returned from Chipumba with a good belt that worked beautifully. At last, no more injuries on the job! Even better, no more getting out of bed in the early mornings to climb the tower. Instead, when the first crow of the cock stirred me from my dreams—which it always did—the steady hum of the spinning machine would sing me back to sleep. But that cock was a persistent one, and often, not even my windmill could guarantee my rest.

"*CHICKEN!*" I screamed. "If you don't shut up, it will be your skinny neck spinning from those blades!"

"*COCK-A-DOODLE-DOOOOOO!!!!*"

It was no use. Conquering darkness on the farm was hard enough, but a noisy chicken—that was impossible.

CHAPTER THIRTEEN

*A*FTER ALL THESE LONG months, I'd hoped that my father's crops would be good enough that we could pay my
back fees and I could return to school. But the famine debts had
been too overwhelming, and as the start of a new school term at
Kachokolo grew closer, we still had no money, not even enough
for tobacco seed or fertilizer. Without a tobacco crop, we had nothing
to sell, which meant we had no money for the remainder of the year. In
fact, it would be several years before we were able to plant another crop of
tobacco.

Instead we began growing crops that didn't need fertilizer and could
be easily sold in the market, things such as soybeans, groundnuts, and
beans. But even though it was great to have these extra commodities to sell,
the prices weren't high enough in the market to make any serious money—
certainly not enough to send me back to school.

One afternoon my father and I were listening to the radio while
working the fields, when an advertisement came on for a local private
school.

"Come to Kaphuka Private School," an announcer said. "Gifted teachers, excellent exam scores, and easy installment plans. Don't waste time!
Come to Kaphuka!"

These schools often bought radio spots, usually torturing me as I sat at home doing nothing. But this time I saw the ad as a good opportunity to raise the question, even though I already knew the answer.

"Papa, what do you think? What about this school? What's happening with my schooling?"

"Well," he said, "we're looking into it. I hope when all these debts are said and done, we'll be able to send you back."

I'm sure it ripped my father apart to get these questions. Not wishing to argue, I accepted his answer and kept on working.

That January, I watched as all my friends returned to class, telling jokes and laughing on the road to Kachokolo. I still saw Gilbert and the others in the trading center for games of *bawo*, and when any of them said things like, "So William, when shall we see you again at school?" or boasted about their good marks on exams, I said nothing, or simply told them, "Please, I'd rather not talk about it." After a while, no one did.

On any given day, you can visit the trading center and see a lot of boys who've dropped out of school and are now doing nothing. Instead of farming or trying to return to school, they're hanging outside the CHiPiKU store in their dirty, tattered clothing, working *ganyu* all day and drinking it away all night. Many of them become only dark shapes through the open door of the Ofesi Boozing Centre, or the zombies who stumble home each morning from the *kachaso* dens.

In Malawi, we say these people are "grooving" through life, just living off small *ganyu* and having no real plan. I started worrying that I would become like them, that one day the windmill project would lose its excitement or become too difficult to maintain, and all my ambitions would fade into the maize rows. Forgetting dreams is easy.

To fight that kind of darkness, I kept returning to the library every week even though I had no idea if or when I'd ever return to school. I kept going so I could increase my general knowledge, and so I would remain inspired. I read all of the library's novels—many of them about the dangers of HIV and AIDS—and spelling books to practice my poor English. And of course, I continued borrowing the books *Explaining Physics, Using En-*

ergy, and *Integrated Science*. Lately, I'd become particularly curious about water pumps, refrigeration systems, and ways to make alternative fuels.

The windmill had been such a success that I began to feel a bit of pressure. I began to see myself like a famous reggae star who'd released a smash album, and now had to produce another hit. Each day at the library, I pored over my texts and tried to come up with my next big idea. The fans were waiting—at least I hoped they were.

Many of those who'd come to see my windmill had suggested similar things: "This looks like a transmitter," or "If you can make this electric wind, you can make a transmitter. That's what it looks like anyway."

This made me curious to see how a transmitter actually worked, and after thinking about it for a while, I went to Geoffrey's house with an idea.

"*Eh*, these people are always saying our windmill is a transmitter, so let's give them what they want."

"What do you mean?"

"Let's build a radio station."

That afternoon, we took out two junker radios from our bag of parts, ones that didn't even have covers attached. First I wanted to test a theory. One night a few weeks before, there'd been a big thunderstorm, and I'd gone inside my room with the radio. I was listening to the Sunday Top Twenty when a huge crack of lightning exploded in the sky. When it did, I heard a blip on my program, as if the lightning had sliced through my signal.

So taking the two radios, I tuned one to a static frequency, then took the second and tuned it to the same place on the dial. When this happened, the second radio went silent: no white noise, nothing. As with the lightning bolt, perhaps the frequency from one radio was penetrating the other? If that was true, surely I could put my own sound on top of that frequency.

One of the radios I was using was a Walkman with a radio and a cassette player. So leaving the first radio tuned to the white noise, I took the Walkman and switched it to tape mode. I saw that wires ran from the

Walkman's tape head to the speakers, so I unhooked them from the speakers and reconnected them to the player's condenser. Because the condenser controlled the frequency, perhaps the cassette music meant for the speakers could instead catch a ride on a frequency wave straight into its fellow radio.

I put my Black Missionaries tape in the deck.

"Here it goes," I said.

I pressed play. Sure enough, the music played loud and clear in the other radio! The Walkman was now my transmitter, meaning that if I had five radios tuned to the same frequency, they'd all be playing the Black Missionaries.

"Now, Mister Geoffrey," I said, "how can I do this with my voice?"

So I unhooked the wires from the condenser and rewired them to a separate headphone speaker, making a microphone. I pressed play and began talking into the mic.

"One two, one two," I said.

I could hear my voice coming from the other player.

"Good afternoon, Malawi. This is your host William Kamkwamba, along with his trusty sidekick, Mister Geoffrey. Your regularly scheduled program has been interrupted."

After that, Geoffrey and I began experimenting with our little radio station. Geoffrey walked outside with the radio, while I stayed in my room singing his favorite Billy Kaunda songs. Even outside, Geoffrey could hear my voice loud and clear. I was really working up a sweat.

"My ears are bleeding!" he yelled. "But please, carry on! This is cool!"

But the farther he got from my bedroom, the weaker the signal became. Beyond three hundred feet, the signal just finally disappeared, which must've been good news for Geoffrey on account of my lousy voice.

"If we only had an amplifier, we could broadcast to distances farther away," I said.

But Geoffrey was scared we would be arrested by the authorities for messing with their frequencies. People were also saying this nonsense about my windmill: "You better be careful or ESCOM power will come arrest you."

If the first people to experiment with great inventions such as radios, generators, or airplanes had been afraid of being arrested, we'd never be enjoying those things today.

"Let them come arrest me," I'd say. "It would be an honor."

SOON I WAS ATTACKING every idea with its own experiment. Over the course of the next year, there was hardly a moment when I wasn't planning or devising some new scheme. And though the windmill and radio transmitter had both been successes, I couldn't say the same for a few other experiments.

The project I was most anxious to get working was a water pump—which had been part of my original idea ever since I'd seen windmills in the book. Although the windmill-driven pump wouldn't come until later, I started work on a prototype pump just to play around with the concept. I modeled it after a picture in *Explaining Physics* of a force pump, which uses a piston and a series of valves to push water through an outlet. The examples illustrated in the book were a car windscreen washer, which I'd never used, and a handheld bicycle pump, which I knew very well.

The shallow well at our house, where we got water for cleaning and bathing, was forty feet deep, so first I needed to find a pipe long enough to reach the bottom. I remembered seeing some pipes in the scrapyard that had once been used for irrigation and were still buried in the ground. So taking my hoe, I went one morning and started digging them up, a process that took two whole days.

The pipes were perfect. The first was a wide PVC pipe, which I could use as my outer barrel. I placed it down the well until I felt it hit the bottom. The second pipe was metal and much thinner, perfect for my piston. Mister Godsten welded a round washer to the end of the metal pipe and left its center hole open. Around the washer I attached a thick piece of bike rubber that would act as the inlet valve, or seal. I then had the welder bend the metal pipe at the top into a ninety-degree angle to create a handle.

When the metal rod was pushed up and down, it created a kind of vacuum inside the plastic pipe. While pulling up on the handle, the water was sucked into the plastic pipe, and when you pushed the handle back down, the rubber seal opened and pressure pushed the water to the surface. The water traveled up the plastic pipe and out a channel hole I'd melted in the side.

But the problem was that the rubber valve created too much friction against the plastic pipe. My sisters, and even women from the next village, had started using the pump, but soon they found it too difficult to operate.

"I can't manage this thing," said my mother. "It feels stuck."

I tried greasing the pipe, but the cold water made the grease as thick as jelly and it didn't spread evenly. I soon gave it up.

The pump wasn't so successful, but my failure in drawing water paled in comparison to my attempt to create biogas.

As I mentioned earlier, deforestation in Malawi has made it very difficult to find firewood for cooking, and gathering wood only adds to this destructive cycle. If there was a good harvest of maize, we usually had enough dried cobs to burn for about four months. But once we'd gone through that, the hunt for wood began.

Each day, my mother or my sisters walked several kilometers to the small blue gum forest near Kachokolo to cut down a bundle of thin trees— a chore that took at least three hours. Most of these trees were alive and green and had to be set aside for nearly five days to dry. This was usually too long to wait, so we burned them anyway, causing thick white smoke to pour from the kitchen windows. Looking inside, I'd see my poor mother stirring the pot of *nsima*, squeezing her eyes closed as tears ran down her cheeks. All the girls in my family developed nasty coughs each year.

In Malawi, every woman has this same burden. And I knew that soon these journeys to find wood would take so long we'd never have our meals. Furthermore, the cycle of deforestation would worsen and create greater problems of drought and flooding. Someone had to help save our women and trees, and I thought, *why not me?*

Ever since I'd built the windmill, several women had asked, "Does electric wind produce enough power for your mother to cook?" Unfortunately, it didn't.

My windmill didn't supply enough voltage to power a proper cooker, so I went in search for other ideas. A few weeks before, I'd been experimenting again with wires and batteries. I'd taken a thick piece of grass—the kind we used to build our roofs and fences—and wrapped it with wire about twenty times. I connected both ends to a twelve-volt cell and felt it heat up. Soon the wire was glowing red hot and the grass caught fire in my hands. It was a simple, kind of childish experiment, but it gave me my next idea.

Okay, I thought, *maybe I can do something with this to boil water.* I couldn't place a metal pot atop a coil of wire because it would only act as a conductor. A clay pot would simply crush the coil. So I fashioned the coil like a magic wand, complete with a plastic handle made from a hollowed-out ballpoint pen. These coils with handles existed already—I'd seen them in the trading center—but they were powered by ESCOM electricity instead of batteries. I hooked a wire to the twelve-volt battery and connected it to the coil, below the handle. I dipped the coil into the water. In about five minutes, it was boiling.

But this was too simple. I had to go bigger. My *Integrated Science* book had a small section on alternative fuels, such as solar power and hydro—both of which I'd studied. But it also mentioned something called biogas, which was made by converting animal waste into liquid fuel, which could then be used for cooking. It explained how the animal waste was buried in a pit. As it heated up over a matter of months, the gas it produced could be tapped through a long valve.

I don't need a pit, I thought. *And I certainly don't need to wait that long.*

Devising my own plan, I snuck into my mother's kitchen and snatched the big, round clay pot she used for making beans. All I needed now was the "organic matter," and I didn't have to look far. Across the compound, Aunt Chrissy kept two goats in a wooden pen behind her house. The ground was covered in their marble-shaped poop. Taking a sugar bag, I made sure nobody was looking and climbed over the fence and into the pen. The goats retreated to the corner and looked at me strangely, but I carried on. I filled the sack until it was spilling over and walked back to the kitchen.

My mother was out working in the garden, which gave me plenty of time and space to work. I dumped the poop into the clay pot and filled it halfway with water, until the brown marbles were swimming around. I then covered the top with plastic *jumbos* and tied a rope around the lip of the bowl, sealing it tight. For my valve, I clipped the top off a radio antenna, creating a hollow tube, then poked it through the center of the plastic. I then corked the top with a reed.

My mother's fire was still warm from breakfast, so I added a handful of maize piths and blew until the fire caught life. I placed the pot in the center and waited for greatness.

About fifteen minutes later, I heard a rumbling inside the pot and the water began to boil. The plastic puffed and danced from the steam, but the rope held tight. My heart began to flutter. I'd give it a few more seconds before starting the final test.

Suddenly, I heard a voice behind me. My mother.

"*What's that smell?*" she shouted.

"Biogas, it's—"

"It's horrible!"

By now the plastic was rumbling like mad, ready to blow. I had to act quickly. It was time to remove the reed and proceed with ignition.

I reached over and quickly popped out the reed, and when I did, a pipe of silver steam came rushing out the top. My mother was right, it smelled vile. I'd set aside a long piece of grass, so I grabbed it now and poked it into the fire, catching a flame.

"Stand back!" I shouted. "This could be dangerous."

"*What?!*"

I stood up and ran to the door, pushing my mother aside. With half my body shielded by the door frame, I stretched out my arm, inching the flame closer and closer.

"Here it goes," I said.

I touched the fire to the piping stream, clinching my eyes to shield them from the flash. But when the flame touched the gas, all it did was sputter out and die. When I opened my eyes, all I saw was a piece of grass, dripping with foul water. My mother was furious.

"Look what you've done; you've ruined my best cooking pot! Boiling goats' poop, I can't believe it. Wait until I tell your father . . ."

I wanted to explain that I'd done it for her sake, but I guessed it wasn't the right time.

In late 2003, while I read books under our mango tree and tinkered with my experiments, my mother went to visit her parents in Salima and stayed for two weeks. Salima is by the lake and very hot, and the mosquitoes there are like small vicious birds. When my mother returned home, she developed high fever and dizziness, then every part of her body began to quiver, as if plunged into an icy bath. We all knew it was malaria.

In sub-Saharan Africa, almost everyone gets malaria at some point. Most people first contract the virus in their childhood. If they don't use proper bed netting to keep from getting mosquito bites, they'll continue to get infected each year until they're old and gray. Because we didn't have nets at my house, malaria was a yearly problem for us all. If we caught the virus in time and took the medicine provided by the clinic, we'd feel better in a week or two. But some stubborn strains of the virus are more ferocious and attack the brain, and they're harder to treat. Every year in Africa, malaria kills over a million people, many of them children.

My mother's symptoms seemed normal, so we put her to bed, and made plans to get her some medicine the following day. But as with my sister Mayless, the fever kept getting worse. By morning my mother was vomiting and shaking even more, and she could no longer speak. Her breathing became so heavy I could hear it across the corridor in my room. By noon, she'd lost the feeling in her legs.

Our village doesn't have an ambulance, so my father hoisted my mother onto his bicycle, told her to hold on, and pushed her to the clinic near the trading center. The nurses took one look at her and demanded that she be rushed immediately to the Anglican hospital in Mtunthama. My father quickly flagged down a pickup and loaded her in the back.

In the waiting room at Mtunthama, the doctor said, "What are your symptoms?"

"I can't really walk," my mother said. "My legs feel paralyzed."

After testing positive for malaria, the doctors gave her two injections in the leg. But there were no free beds at the hospital, so the doctor sent her home.

Two days later, my mother began slipping into a coma.

The morning it happened, we managed to get her to her feet and walk her outside. We lifted her onto the bicycle, doing our best to keep her from falling off the saddle.

"Mama, you have to hold on," I said, but it was no use.

As we pushed her down the trail, her limp body kept collapsing like a sack of beans. Her head rolled back, so I grabbed a handful of hair and held her up.

"Don't worry," my father kept telling my mother, the fear straining his voice. "Just try to hold on until we get to the road. We're taking you to the hospital now. They'll fix you and make you better."

The pickup stop was just a small space under some mango trees near the clinic and primary school. Once there, we gently pulled my mother off the bike and laid her on the grass. In minutes, a pickup came rumbling from the trading center, headed for Kasungu. My father waved it down.

"Make way!" he screamed. "My wife is sick!"

About ten people were squeezed in the back of the truck, along with crates of empty Coke bottles and a few sacks of maize. When the passengers saw my mother, several of them jumped out and made room.

"It's bad malaria!" my father shouted to the driver. "Take us to Mtunthama!"

We slid her into the bed of the truck and leaned her back against the cab. My father sat beside her, holding up her body and placing her head against his shoulder. The road to Mtunthama was filled with holes and bumps, and laying her flat would only rattle her body against the bed.

"Look after your sisters, William!" my father said. "Tell my sister Chrissy and the others where we've gone." And like that, the truck sped away.

The truck reached the hospital in fifteen minutes. Once there, my father carried my mother in his arms through the door.

"We need to see the doctor now!" he shouted.

My mother was quickly admitted into a room, where the doctors administered an IV drip to battle the virus.

"This doesn't look promising," the doctor said. "It appears it's gone to her brain."

The room had pink walls and ESCOM-powered lights. Various posters on the walls showed people suffering different diseases, things like AIDS, tuberculosis, and gonorrhea. Another woman lay in a bed beside my mother. She was from Chamama and kept vomiting into a bag.

That afternoon, my aunt Chrissy and Socrates' wife, Mary, arrived and stayed through the night, keeping vigil. My father came home to try and sell some maize and soybeans to pay the medical fees. He was quiet and kept pacing the courtyard, as if he was waiting for something.

"Papa, is Mama going to be okay?" Mayless asked.

"She's very sick. Pray for your mother."

My father was able to sell a few kilos of grain the next morning in the trading center and he immediately returned to the hospital. I volunteered to stay home and watch my sisters; I was also too afraid to see my mother in such a state. The next day, when I finally did get the courage to go, I quickly regretted it.

My mother's dark skin looked drained of color against the white sheets, and her lips were dry and cracked. Her chest rose and fell from her labored breaths like a toy boat on the waves. Her eyes were closed, but her eyeballs were dancing inside.

My mother later told me that deep in the recesses of her darkness, she'd already accepted death. She'd given up her fight and was waiting for Jesus to come and take her. But something wouldn't allow her to depart. She'd feel her body sinking through the bed, only to rise up again. When this happened, her eyes opened and she saw people she knew, standing above her. Everything would go dark again, and the cycle would repeat. Seeing all those people made her remember her children. At one point in her darkness, she saw a vision of Tiyamike, so young and still so fragile. In the dream, she was alone and frightened because her mother was dead. Thinking of her daughter, my mother struggled to free herself from the blackness for good. It was a terrible fight, which explained why her eyes kept squirming like termites. When she managed to break free and open her eyes, she saw me standing there.

"*Tiyamike!*" she screamed. "*Where's Tiyamike? Where's my baby?*"

I jumped back, as if she were a snake. Her eyes were wide, and pulsing with fear.

"*Tiyamike!*"

"Tiyamike's at home," my aunt said. "You'll see her soon. Don't worry."

Slowly, as if being deflated of air, my mother then slipped back into the darkness. Every time she came back, she'd scream my sister's name again. Seeing your mother like this is like having God steal the sky from over your head. I was certain she was going to die, and witnessing this made it even more emotional. But after several days, her fever miraculously broke. I'd never prayed so hard in my life.

NOT LONG AFTER MY mother came home, Gilbert told me that his father wasn't doing so well. Ever since the beating by the president's thugs, the chief had lived in fear for his life, and his health had grown worse.

Whenever I visited, he always appeared quiet and weak, and lately, he'd lost a lot of weight. I saw him sleeping on the sofa or taking walks alone in his fields when the sun was warm. But because he was the chief, I never really spoke to him. It wasn't my place as a young man.

A few months later, when I was visiting Geoffrey at the maize mill in Chipumba, we passed some women on the road who said they had some bad news.

"Brothers, our Chief Wimbe is no more," they said, tears in their eyes.

Geoffrey and I got on my bike and tried to hurry home, but my tire burst and I had to push. While we were struggling with the bike, preparations for the funeral had already begun. A stream of cars and trucks rumbled past. The lorries were filled with chickens and goats and great bags of flour to feed all the guests, who were already arriving to honor the chief. Dozens of village women were also on their way to cook for mourners. All of these people were silent as they passed. Not one radio could be heard. Once the traffic had cleared, I heard the great pounding of a drum, like none I'd ever heard—a dull boom, like a hammer banging the very shell of the sky. A chief was dead.

Hundreds of people were gathered around Gilbert's house when we arrived. I saw my parents and sisters and aunts and uncles, along with all the traders from the market. Women ran back and forth carrying buckets of water on their heads. Others hunched over fires, wrapped in wood smoke and sweating as they stirred giant pots of *nsima*. A church choir stood under the blue gums softly singing "The World Is Not My Home" while a steady line of people poured out of Gilbert's front door, wailing and screaming.

"Our king has left us!" they cried. "What will we do?"

Geoffrey and I sat down under a tree and waited. Soon, someone came and said Gilbert was ready to see us. He was sitting on the other side of the house under some trees. My friend looked like he was in shock, even though he was trying to stay strong for his mother. Seeing him filled me with sadness.

"I heard about your father," I said. "You know I'm here for you in this terrible time. God is in control." I didn't know what else to say, so I just remained silent and comforted my friend.

The funeral ceremony was held in the yard of the primary school underneath the big blue gums. A light rain started falling. Someone erected a large canopy for the family and delegation of chiefs and officials who'd arrived from all over. They crowded under the shelter while hundreds of villagers huddled outside, moaning and sobbing in the downpour. Up front, the chief's body lay in a closed wooden coffin, covered in wildflowers. Every church and mosque around Wimbe was in attendance. Their choirs took turns huddled around the wooden box singing songs in Chichewa. When they finished, the silence was broken by the banging of the drum.

The funeral dirge was slow and steady, then rose in tempo. Atop a fast and furious rhythm, out walked the Gule Walmkulu. More than fifty of them pressed around the coffin, each wearing black masks in the shape of cows' heads, with long exaggerated snouts, black horns, and round bulging eyes. The mystical dancers had been there for days, camping behind Gilbert's house and huddled around their fires, never revealing their faces. Now several of them broke free of the group. Their bodies locked in spasms as they began to dance. They crouched low in unison and kicked out their legs, sweeping their arms over the red soil as if telling the earth it was about to receive our chief.

After the dancing, the funeral procession filed down to the graveyard near the Catholic church. The chief's grave had been dug much like Uncle John's, with a smaller compartment at the base of the pit. Prayers were said and wreaths were laid on the coffin. Mister Ngwata, the chief's messenger, appeared in his khaki policeman's uniform and fired a shotgun into the air. With our ears ringing and faces stinging from tears, we watched as our beloved leader was lowered into the ground.

IF THE DEATH OF our chief wasn't bad enough for our district, later that year, another famine fell upon the country. It arrived in spite of some newfound hope: in May 2004, our despised President Muluzi had finally stepped down and made way for new elections. Malawians then chose Bingu wa Mutharika as our new president. Mutharika was a respected man who'd earned degrees in economics in the United States and held high posts in the United Nations. President Mutharika pledged that change would soon come to Malawi, and one of the first things he did was reach out to us farmers. For the next planting season, his government began subsidizing fertilizer, and that meant my family could afford it for the first time in three years.

Fertilizer coupons reduced the price from four thousand kwacha to nine hundred fifty, with each family getting four coupons each. However, this scheme didn't stand a chance against the culture of corruption that already existed. Instead of distributing the coupons to farmers, many local leaders horded them and sold them to the people with the most cash.

In December 2005, each farming family received their four coupons. To split up the work of hauling these heavy bags, my father and I each took two, then waited in line all day at the ADMARC in Wimbe, where the government sold its maize and fertilizer. My father received his bags and left, but when it was time to get mine, the corrupt officials had given most of them away to their friends. They closed the shop early and the farmers in line nearly began to riot.

I ran up to the window so I could look inside to see what was happening, and just as I pressed my face to the glass, I felt a sharp pain in my back. One of the government security thugs was standing behind me, beating me with a hose pipe.

"Get out of here, you little monkey!" he yelled, swinging the hose like mad. He struck me again on the arms, then several times in the back. Pain shot down my spine, and I ran as fast as possible. Other guys who'd been standing nearby were caught against the wall and whipped mercilessly. Once I was safely away, I felt my face flush with anger. I wanted to kill him, but he was quite muscular and I didn't stand a chance.

"You're lucky my father has already gone!" I shouted.

We later managed to get a couple more bags of fertilizer and prepared to plant our crop. At first, the rains came like normal. We planted our seeds, going down the rows with our fertilizer, dumping a spoonful in a hole beside the seed station and covering it back with soil.

By January the seedlings had begun to sprout and were ankle high and showing their little arms, so happy to be feeding on such delicious rainwater and fertilizer. But just about the time they reached my father's knees, the rains stopped completely. The sun rose up every morning angry and blistered the poor seedlings, causing them to wither and crouch. Soon their leaves were dry and brittle. If we'd put a match to them, they'd have burned to a crisp.

"Next year there'll be trouble," my father said.

"I don't know if I can take another," I answered.

It rained only a few more times in February, enough for many of the plants to produce a pith. But come *dowe* season, most of the ears were stunted and deformed. The harvest would be terrible. The government quickly promised intervention, but in the meantime, people grew angry and scared.

During the famine of 2002, the people of our area blamed the unpopular Muluzi government, venting their rage on the corrupt officials who'd sold our maize surplus. But this time, instead of acknowledging the cruel weather, they began blaming magic. And that meant blaming me.

Superstitions were still very strong throughout the country, and several incidents in the news had stirred these fears even more. The previous famine had led to reports in the southern region that the government was banding with packs of vampires to steal people's blood, then selling it to international aid groups. Mobs had gone crazy. One old man was stoned to death and three Catholic priests were beaten. The government denied having any involvement with vampires, but that didn't seem to make these rumors go away. "No government can go about sucking the blood of its own people," President Muluzi had said at the time. "That's thuggery."

Following this vampire episode, a strange beast appeared in Dowa and began attacking villages. Some said the beast looked like a hyena, others said it was a lion with the face of a dog. Three people were mauled to death, and sixteen others had their hands and legs ripped from their bodies. The attacks caused thousands to flee their homes and sleep in the forest, where they were even more vulnerable to this alien creature.

The police conducted all-night searches, and one evening, they managed to corner the beast in a thicket. But according to the newspapers, whenever the police fired their rifles, the beast split into three separate animals and disappeared into the bush. Villagers then visited the *sing'anga*, who concocted a powerful potion and flung it into the trees. The next morning, the beast lay dead on the road—its corpse no bigger than a dog. When the elders tried to burn its body, no fire would consume it.

The villagers returned to their homes. However, just when everyone thought they were safe, a second beast began attacking and killing again, sending thousands back into the bush. It was later concluded that the beast was the product of magic. A certain trader near Dowa had purchased some thunder and lightning from another powerful wizard and later refused to pay for it. In retaliation, the magic man had sent the beast against his family. All those who had been mangled and killed were his relatives.

Following the strange beast of Dowa, many people across Malawi reported having their private parts stolen in the night, many of them waking up in the morning with their sheets bloody. Men who'd been drinking in bars were the easiest targets. As they stumbled home in the darkness, an evil creature—perhaps a gang of witch children—would pull them behind a tree and remove their parts with a knife. It was later revealed that most of the victims had been virgins, and their parts had been sold to witches, Satan worshippers, and business tycoons.

This was all so worrisome that the opposition leader, the Honorable John Tembo, addressed the problem in parliament, saying something like, "It's not acceptable to sell other people's private parts, especially while leaving your own."

This wave of superstition and fear soon came to Wimbe. People said some witches were living near the trading center and using children for magic. One night the witches instructed the children to attack an old man who was known as a good Christian. While the old man was sleeping, the children magically removed his head and used it for soccer. (This often happens while we sleep—the witch children can take our heads and return them before morning, all without us knowing. It's a serious problem.) This soccer game was no ordinary match, but a great tournament in Zambia among all the children of the devil. The trophy cup was filled with human meat, which was to be eaten on Christmas Day.

The Malawian team was matched against the young witches of Tanzania. But as the game commenced and crowds hissed from the stands, something terrible happened. The ball deflated. When this happens, the head cannot be returned and so the old man died. After a new head was introduced, one of the Malawian children touched the ball with his hands at the eighteen-yard post, and that led to a penalty kick. The Tanzanians won the match 1-0.

In the witch plane back to Wimbe, the other children mocked and cursed the young boy.

"Why did you hand the ball?" they shouted. "We would've won!"

The other witches then beat the boy terribly with their magic. They were most upset about not getting the trophy cup of meat. The boy's grandfather happened to be a traditional chief in the neighboring village, so the witches gave the boy a choice.

"Tonight you must kill your grandpa and bring us his meat, otherwise we'll eat you instead."

The boy was so badly beaten by their magic that he awoke the next morning too exhausted to move. When his father came to rouse him from bed, the boy confessed everything. He explained the tournament, how the man's head had deflated, and the threats by the other witches.

"An old man is dead in the village," he said. "And we are the ones who killed him. Now they want me to kill our grandfather."

The parents became angry and reported the matter to their grandfather, the village headman. The child gave them the name of the witch who'd recruited him and the other children, and a mob was sent to his house. The man was beaten with clubs and sticks and barely escaped with his life. The boy's parents then reported the matter to Chief Wimbe (Gilbert's cousin, who'd since taken the throne). The chief investigated and made three arrests. These witches were convicted in the traditional court and made to pay a large amount of money. Sadly, our country's constitution doesn't have a clause that protects us from witchcraft. Because it's so difficult to prove, the authorities are limited in their investigations. They can eventually convict a wizard of violating the rights of a child, but never for kidnapping or murder. Hopefully one day that will change.

ANYWAY, THESE INCIDENTS ONLY heightened the fears and belief in evil powers. So in 2006, when the maize failed to grow and the likelihood of famine was strong, people started blaming magic. One afternoon in March, when the fields were withering in the sun, giant storm clouds began building in the distance.

"Look, the clouds are gathering," people said. "Today we'll have rain!"

"Yes, finally we're saved!"

We'd gone weeks without rain and the sight of thick black thunderheads was something to celebrate. But just as these clouds began to pass overhead, a strong wind began to blow. It whipped up the red soil and threw it in our eyes and mouths, sending miniature cyclones tearing through the fields and across the courtyards. Little by little, this wind blew the clouds away.

With nothing but the blazing sun left in the sky, a few people gathered at my house and pointed up to my windmill. The blades were spinning so fast the tower rocked and swayed.

"Look, this giant fan has blown away the clouds. His machine is chasing away our rain!"

"This machine is evil!"

"It's not a machine—it's a witch tower. This boy is calling witches."

"Wait!" I screamed. "The drought is all over the country. It's not just here. Electric wind is not the cause."

"But we saw it with our own eyes!" they said.

I became very afraid these people would collect a mob and tear down my windmill, or worse. I kept a low profile for the next week, stayed inside, and even stopped the blades during the day so they wouldn't raise more suspicions. At the trading center, people approached Gilbert.

"You can tell us the truth," said one trader. "Is he really a witch? Is it true what he says about this electric wind?"

"He isn't a witch," Gilbert answered. "It's a windmill, a scientific machine. I helped him build it."

"Are you sure?"

"I'm sure. Go see for yourself."

Many of them knew what my windmill was actually for, and a lot of them had even stood in the line and charged their mobile phones. But pointing the blame at me helped them get over their fears about the upcoming famine. Luckily, not long after that, the government stepped in and released tons of maize on the market. A few months later, some aid agencies arrived and offered further assistance. No one starved or died. A catastrophe had been avoided, but still, it revealed the kind of backwardness in our people that really frustrates me.

SADLY, MAGIC WAS OFTEN the scapegoat for another great tragedy in Malawi: the spread of HIV and AIDS. Around this period, about 20 percent of all Malawians were infected, and many thousands died each year. Not only had it killed tens of thousands of our teachers and further deprived our students of a good education, by 2008, AIDS had also killed many of our national entertainers, including many musicians, robbing our country of one of its proudest treasures.

People died from AIDS because of stubbornness and lack of education. For many years, our villages didn't have proper clinics that addressed HIV because of the stigmas attached to it. People weren't taught to use protection during sexual relations, and once they became sick, many fell into denial. Others visited magic men who recognized the HIV symptoms right away, but still told the patient, "You're right, brother, you've been bewitched. Luckily, I have just the thing."

These wizards also claimed to treat other serious ailments requiring medical attention, which led to many senseless deaths in our country. One of the biggest was diarrhea. A person would see the wizard and complain of serious pains in his or her stomach.

"Oh, I know what's wrong," the wizard says. "You have a snail."

"A snail?"

"I'm almost positive. Do you feel movement in your intestines?"

"Of course, the pain is terrible."

"That's a snail. We must remove it at once!"

"Then do it quick, it hurts so bad."

So the wizard goes into his bag of roots, powders, and bones and pulls out a lightbulb.

"Lift up your shirt."

Without plugging the lightbulb into anything, the wizard begins moving it slowly across the person's stomach, as if to illuminate something only he can detect.

"There it is! Can you see the snail moving?"

"Huh?"

"Can't you see it moving its antennae? That explains why you're crying so much."

"Oh yes, I think I can see it. Yes, there it is!"

The wizard goes into his bag again and pulls out a bundle of dried roots, which he dips into a jar of water. After some minutes, he splashes this magic potion onto the person's stomach.

"The moment this water comes in contact with the snail, it will die and come out."

After a few minutes, he looks up.

"All better?"

"Yes, I think the snail is gone. I don't feel it moving."

"Good. That will be three thousand kwacha."

THERE WAS SUCH A stigma attached to AIDS that most people who suffered from it didn't seek any medical help beyond what they could get from the *sing'anga*. You'd see them in the trading center looking very thin, their hair turning yellow and pale, and the color leaving their eyes. Sometimes you'd see them being loaded in the backs of pickups bound for Kasungu hospital, never to return.

This lack of education also led to harsh teasing and discrimination by the population. Often young kids would harass people on the roads who appeared thin and weak and showed the signs of being *wakachilombo*—a person infected by a virus.

"Look at the *wakachilombo* with AIDS!" they'd shout. "He's gonna die! Mister, you'd better prepare your grave!"

One afternoon while playing *bawo* in the trading center, some health personnel from Wimbe clinic approached us boys and began chatting. They said they were starting a youth club to encourage people to get tested.

"Why don't you boys join us and sensitize your community," they said. "We need help spreading the truth."

That day, the Wimbe Youth Friendly Health Services Club was born, becoming one of my most cherished activities. We began meeting every Monday and learned about prevention and treatment of HIV and how to approach others about the subject. The class was filled with other kids our age, most of whom had also dropped out of school due to lack of funds. Geoffrey and Gilbert were also members. I was so happy to be in a classroom-type environment again where I could learn and socialize with others, where we could show our smarts and joke around with our pals. Just as the windmill and scrapyard had done, the classes worked to replace school in my mind and kept me focused. I put everything into my club.

The doctors at the clinic were so impressed by our enthusiasm they asked me to write a play to help persuade people to get tested. Over several days, I prepared a great production—more in my mind than on paper— which I titled: *Maonekedwe apusitsa*. Or simply: *Don't Judge the Book by Its Cover.*

The play was performed two weeks later in the trading center. Gilbert and I put up posters announcing the show, and even asked our new chief—Gilbert's cousin—to summon a few Gule Wamkulu to help draw the crowds. The morning of the play, we got our gang of actors together and marched through the trading center, shouting, "Come one, come all!! See the HIV drama! See the acrobatics of Gule Wamkulu!" We set up in the market center and soon, about five hundred people had gathered around. Many of the shops even closed during the performance.

The play involved a certain married couple in the capital Lilongwe (played by my friends Christopher and Mese). The husband sends his wife to the village to grow some maize. Well, as you know, life can be very difficult in the villages. People get hungry, the work is hard, and the sun is very hot. Not accustomed to this kind of farm life, the wife loses a lot of weight. When she returns home, the husband becomes angry.

"Why are you so thin?" he asks.

"The village is tough," she says.

"You're lying. I think you've been sleeping with other men and now you have AIDS."

Little did the wife know, but while she was gone, her husband had been spending every night in the bars of Area 47, where he'd slept with many prostitutes.

The wife pleads her case with the husband, denying his accusations. But the husband doesn't accept her story. "Go back to the village," he tells her. "If you want to sleep around, then go and do it. Get out of my house!"

Luckily, a friend arrives just in time and sees them fighting. (This was my part.)

"Wait a minute, brother, what's the problem?" he says.

"I sent this woman to the village to grow maize and look how thin she became. She says the farming was hard work, and there was no food in the village. But I say she has AIDS, and I'm not staying with her anymore!"

The friend says, "Brother, you can't tell if someone has AIDS by just looking at them. It could be hunger, tuberculosis, or many other things. The only way to know is to be tested at the local Volunteer Counseling and Testing center."

"Fine!" says the husband. "We'll go down there and prove this once and for all."

The husband thinks because he's so muscular, all those prostitutes in Area 47 haven't hurt him. But after being tested at the VCT center, sure enough, the doctor (played by Gilbert) informs the husband he's HIV positive. His wife is negative.

"You're cheating me!" he tells the doctor. "This can't be true!"

"I wouldn't tell you a lie," the doctor says. "But don't worry. This isn't the end. You can still live your life if you play by a few simple rules."

The wife becomes afraid. "There's no way I can stay with you. You're right, husband. I should leave."

"Don't be foolish," says the doctor. "You can stay together. Just be careful and use protective measures."

The wife is so relieved and gives her husband a hug. "I still love you," she says. "I'll stay with you until death do us part."

When the play was over, the crowd shouted and cheered and threw handfuls of paper in the air. Once we actors cleared the stage, the Gule Wamkulu closed the show with an electrifying grand finale. I won't say our little play inspired loads of people to be tested right away, but it did help change attitudes. These days, with the help of government programs and new VCT centers, more people are being educated and tested and AIDS isn't such a taboo. Even the magic men started sending their clients to the clinics, preaching science and medicine over their charms.

My popularity as an inventor and activist soon attracted other opportunities. Not long after the play, one of the teachers at Wimbe Primary asked if I'd be interested in starting a science club for the students. He'd been impressed by my windmill and wanted me to build one on campus.

"The students look up to you," he said. "Your skills in science will really challenge their brains."

"Sure," I said. "I'll do it."

The windmill I created at the school was small, much like my first radio experiment, with blades made from a metal maize pail and a radio motor for a generator. I attached it to a blue gum pole and ran the wires into an old Panasonic two-battery radio that I'd repaired. I did this during recess one morning when all the kids were in the grove playing soccer. As I worked, a crowd gathered around to watch. When I connected the wires, the music blasted through the schoolyard.

"Keep quiet. I want to listen!" they screamed.

"Don't push me!"

"Let me see!"

The windmill not only allowed students to listen to music and news, but they could also charge their parents' mobile phones. Each Monday, I told them about the basics of science and explained the importance of innovation, like how ink was first made by using charcoal. I also demonstrated the cup-and-string experiment featured in my books, to help explain how a telephone works.

I walked them through the steps of how I'd built everything using simple scrapyard materials, and I hoped I'd inspire them to build something themselves. *If I can teach my neighbors how to build windmills,* I thought, *what else can we build together?*

"In science we invent and create," I said. "We make new things that can benefit our situation. If we can all invent something and put it to work, we can change Malawi."

I later found out that some of the students had been so inspired by the windmill in the schoolyard, they went home and made toy windmills themselves.

I began to imagine what it would be like if all of those pinwheels had been real, if every home and shop in the trading center each had a spinning machine to catch the wind above the rooftops. At night, the entire valley would sparkle with light like a clear, starry sky. More and more, bringing electricity to my people no longer seemed like a madman's dream.

CHAPTER FOURTEEN

*I*N EARLY NOVEMBER 2006, some officials from the Malawi Teacher Training Activity were inspecting the library at Wimbe Primary when they noticed my windmill in the schoolyard. They asked Mrs. Sikelo, the librarian, who'd built it, and she gave them my name. One of them telephoned the head office of MTTA in Zomba, in the southern region, and told his boss, Dr. Hartford Mchazime, what he'd seen.

A few days later, Dr. Mchazime drove five hours to Wimbe. His driver took him to our house, where he asked my father if he could speak to the boy who'd built the windmill.

"He's here," my father said, and called me from my room.

Dr. Mchazime was an older man with gray hair and kind, gentle eyes. But when he spoke, his command of language was large and powerful. I'd never heard anyone speak such good Chichewa, and when he spoke English, it was simply inspiring.

"Tell me everything," he said, asking me about the windmill and how it had happened.

I explained my windmill, as I'd done hundreds of times before, then took him through our house and demonstrated how my switches and the circuit breaker worked. He listened carefully, nodding his head, and asked specific questions.

"These are very tiny bulbs," he said. "Why aren't you using big ones?

"Big bulbs require AC power," I said. "In order to have many lights, I have to charge a battery, which is DC. These tiny car lights are the only DC bulbs I could find."

"How far did you go with your education?"

"I dropped the first year of secondary school."

"Then how did you know this stuff about voltage and power?"

"I've been borrowing books from your library."

"Who teaches you this stuff? Who helps you?"

"No one," I said. "I've been reading and doing it alone."

Dr. Mchazime then went to see my mother and father.

"You have lights in your house because of your son," he said. "What do you think of this?"

"We're proud," my mother said. "But we thought he was going mad."

Dr. Mchazime laughed and shook his head.

"I want to tell you something," he said. "You may not realize, but your son has done an amazing thing, and this is only the beginning. You're going to see a lot more people coming here to see William Kamkwamba. I have a feeling this boy will go far. I want you to be ready."

The visit left me a little confused and very excited. No one had ever asked me such questions before, and no one had taken that kind of interest. That afternoon, Dr. Mchazime returned to Zomba and told his colleagues what he'd seen.

"This is fantastic," they said. "The whole world needs to know about this boy."

"I agree," Dr. Mchazime said. "And I have just the idea."

The next week, Dr. Mchazime returned to my house with a journalist from Radio One. It was the famous Everson Maseya, whose voice I'd heard for years, and here he was at my house to interview me.

"What do you call this thing?" he asked.

"I'm calling it electric wind."

"But how does it work?"

"The blades spin and generate power from a dynamo."

"And in the future, what do you want to do with this?"

"I want to reach every village in Malawi so people can have lights and water."

Two days later, while waiting for the Radio One interview to air, Dr. Mchazime came again with even more reporters. These men represented all the great media organizations in Malawi: Mudziwithu and Zodiac Radio channels, the *Daily Times,* the *Nation, Malawi News,* and *Guardian News.* They poured out of the car with their cameras and tape recorders and flocked around the windmill.

For two hours, they moved through the house, elbowing and shoving one another to get the best pictures of my switches and battery system.

"You've had your time. Now it's my turn!" they shouted.

"Move aside, my paper is bigger!"

Soon our yard was filled with crowds from the trading center who'd gathered around to gawk at the famous journalists who'd come to our village.

Journalists visit my village to write about my windmill. To us villagers, these men were like celebrities.

WILLIAM KAMKWAMBA AND BRYAN MEALER

"Look, it's Noel Mkubwi from Zodiac!" they said.

"Finally we see his face. What a handsome man!"

"Look, he's interviewing William!"

One of the reporters even climbed my tower and studied the blades and pulley system, taking pictures the whole time.

"Mchazime, this chap is a genius," he shouted.

"Yes," he answered, "and this is the problem with our system. We lose talent like this all the time as a result of poverty. And when we do send them back to school, it's not a good education. I'm bringing you here because I want the world to see what this boy has done, and I want them to help."

Dr. Mchazime told us that as a boy he had endured his own setbacks with education. His father had also been a poor farmer who'd struggled to feed and clothe his family. But his father had learned the value of an education. While working in the gold mines of Rhodesia, he'd been denied several opportunities because he'd never gone to school. That failure seemed to haunt him the rest of his life.

At one point when Dr. Mchazime was young, his family barely had enough to eat. He had volunteered to drop out of school and work so his brothers could go instead. His father refused, saying, "All of my kids will stay in school. I'll do whatever it takes." It took nearly ten years for Dr. Mchazime to complete his secondary education. When he was thirty-three years old, he was finally accepted with a scholarship to the University of Malawi in Zomba, later earning master's and doctorate degrees from universities in America, Britain, and South Africa. Before working for the MTTA, he'd written many Malawian textbooks, including my own Standard Eight English reader.

THE DAY AFTER THE journalists came to visit, the interview with Everson Maseya finally aired on Radio One. I was behind the house chatting with my aunt, when my mother shouted, "William, come quick. It's coming on!"

My family all gathered around the radio, and I heard the announcer

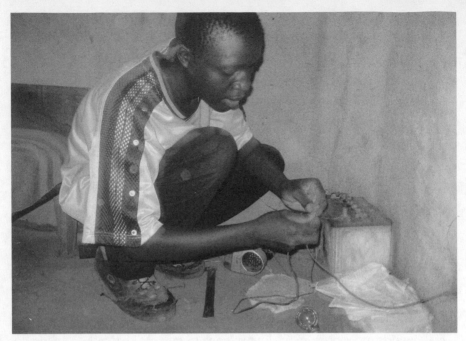

Here I am connecting the windmill to the battery for the journalists, trying not to burst out laughing. I was so nervous, but even more excited.

say, "A boy in Wimbe near Kasungu has made electric wind." When my voice started to come through the speakers, my sisters began to cheer.

If the radio show wasn't enough good fortune, the story in the *Daily Times* was published the next week, with a big headline that said, "School Dropout with a Streak of Genius." The story has a photo of me pretending to connect the wires to the battery in my room, while trying not to smile. That afternoon, I took the paper to the trading center to show everyone what the madman had done.

"We also heard you on the radio," people said. "Did you have to go to Blantyre?"

"No, they came to me," I answered.

"Really? We're very proud. You represented us well, and we're so impressed at how well you spoke."

In a way, it took having these reporters come to my house to make our town finally accept my windmill. I don't know, but I think it was a kind of validation. After the radio and newspaper coverage, the number of visitors to my house increased tenfold.

Shortly after the story ran, I started some much-needed improvements on the windmill. I'd realized the big acacia tree behind the latrine was blocking my strongest wind and I needed to go higher. So with the *Daily Times* story under his arm, my father was able to convince the manager of the tobacco estate to part with several giant poles, which I used to build a tower that was thirty-six feet high. Once I moved it away from the tree, the speed of my blades doubled, and so did the voltage.

AFTER VISITING MY HOUSE with the reporters, Dr. Mchazime went back to Zomba and gathered his colleagues.

"I think this chap should be sent back to school," he said. "He needs to continue his education and develop his abilities. That way these inventions will be credible and people will respect what he's doing. Without education, he's limited."

"We agree," one of his associates said. "Perhaps we can find an organization that can support him."

"Eventually, yes," Dr. Mchazime said. "But we need to put him in school as soon as possible. In this office, can we contribute something for his fees?"

Dr. Mchazime pulled out a wad of kwacha from his pocket and tossed it on the table. "Look, here's my contribution, my own pay. Who will follow?"

By the end of the day, Dr. Mchazime had collected nearly two thousand kwacha.

That week, Dr. Mchazime contacted the Ministry of Education and asked them to find a good school for me to attend. No one answered his calls or letters, so he drove to the office of the head of secondary education.

"I sent a letter," he told the official.

"We received your letter," she said. "This boy has a very interesting story. We'll find a place for him, but it can't be now."

"You people are delaying him," Dr. Mchazime said. "He's growing up, and the more you delay, the more schools will say he's too old. Try to do this faster."

The woman said they'd be in touch, but no one called back. Dr. Mchazime returned and was told to go to Kasungu, to see the division manager of central east schools. Dr. Mchazime got in his car and drove four hours to her office.

"I've read this kid's story," the woman said. "He's very interesting."

"Of course he's interesting, so don't waste his time," Dr. Mchazime said. "He needs to be in school immediately."

"There are procedures that must be followed," she said.

"Surely you can make an exception. Surely there's some kind of waiver?"

"Okay," she said. "I'll go see this windmill myself."

While I was in the trading center running errands for my father, the manager of schools came to see my windmill, along with several people from the Ministry of Labor. They wore suits and their faces dripped with sweat as they stood in the hot sun. They didn't explain why they were there—they just asked my mother if they could look around. An official from Labor then said to his colleagues, "This boy has special talent. We need to take him in the system. We need people like this in government!"

When the manager of schools returned to her office, she called Dr. Mchazime.

"You're right," she told him. "This boy needs to be in school, and we have just the place."

"It must be a boarding school, one that specializes in science. Please don't delay."

"We'll take it from here," she said.

WHILE THIS WAS HAPPENING, something else amazing was taking place without my knowledge. The day after the *Daily Times* article ran, a Malawian in Lilongwe named Soyapi Mumba brought the article to his office. Soyapi worked as a software engineer and coder at Baobab Health, an American-run NGO that's working to computerize Malawi's health care system, which is very unorganized with its stacks of old record books. One of Soyapi's bosses, a tall American named Mike McKay, liked the article about my windmill so much that he wrote about me on his blog Hacktivate. That blog entry caught the attention of Emeka Okafor, a famous Nigerian blogger and entrepreneur, who was program director of a big conference called TEDGlobal 2007.

Well, Emeka wanted me to apply to be an official "fellow" at this conference, and for three weeks, tried very hard to find me. After harassing the reporters at the paper every day, he finally tracked down Dr. Mchazime. In mid-December 2006, Dr. Mchazime came to my home with the TED application. We sat down under the mango tree, and he helped me fill in the questions, plus write a small essay about my life. When he left, I still had no idea what TED was, or what it even meant (TED means Technology, Entertainment, and Design, and it's an annual meeting where scientists, inventors, and innovators with big ideas get together and share).

I wasn't even sure what a conference was, or what people did at such things. The application didn't say where it was held. I suspected Lilongwe but didn't know. I began to imagine myself walking the streets of the capital, seeing all the many kinds of people. People said Lilongwe was filled with thieves, but I wasn't afraid. If anything happened to me, I'd already decided I'd go to the market and ask help from some women. Women will always help you. But what would I wear to this conference? Everything I owned hung from a rope in my bedroom and was red from the roof dust. Even still, it gave me something to dream about.

In the beginning of January, just after New Year's, one of Dr. Mchazime's colleagues called Geoffrey's phone (I didn't have a mobile), to pass along a message that I'd been selected for the TED conference.

"Tell him to get ready," he told Geoffrey. "He's going on a journey."

Geoffrey didn't know the details, but said Dr. Mchazime would phone me. Later that week, Dr. Mchazime called again. I happened to be standing nearby and Geoffrey handed me the phone.

"You're going to Arusha, Tanzania," he said. "You'll be honored with other scientists and inventors. People from all over the world will be there. Perhaps something good can come from it."

Wow, Arusha. I imagined the bus ride. How long would that take? I'd have to bring plenty of food, perhaps cakes and roasted maize. After all, I had no money.

"One important thing," he said. "We should book your flight before it fills up."

"I'm traveling by plane? *My God.*"

"Yes, and they wish to know if you want a smoking or nonsmoking room in the hotel."

"Hotel? I'm staying at a hotel?" I thought for sure I'd be sleeping in one of those guesthouses near the boozing dens where the poor people stay.

"Of course you're staying in a hotel," he said. "And I have other good news. William, you're going back to school."

FINALLY, AFTER MONTHS OF haggling with the Ministry of Education, I was given permission to attend Madisi Secondary, a public boarding school an hour from my home. It wasn't one of the science-oriented schools that Dr. Mchazime had suggested. The headmasters at those schools weren't willing to accept me on account of my old age and number of years I'd been a dropout. However, the headmaster of Madisi, Mister Rhonex Banda, was so moved by my story he offered to spend the extra time helping me catch up. I was so terribly behind.

While Dr. Mchazime planned my trip to Arusha, I packed my things and went to school. This was the first time I'd ever lived away from home. I had a black suitcase I'd purchased a few weeks earlier in Lilongwe while visiting Geoffrey's brother Jeremiah, who'd moved to the city a couple years before. Along with a toothbrush and toothpaste, I packed my flipflops, a blanket, three T-shirts, a pair of trousers, one nice shirt, one pair of socks, and two pairs of underpants. The bag had wheels, so I rolled it out through the courtyard and stopped under the mango tree. My parents were waiting, along with Geoffrey.

"I guess I'll see you soon," I said.

"Work hard," my father said. "I want you to know we're very proud."

Geoffrey strapped my suitcase to his bicycle, and we rolled it down the trail toward the pickup stop. Along the way, we stopped at Gilbert's house.

"We don't have phones, how will we talk?" Gilbert asked.

"It will be difficult," I said.

"Maybe I can come visit you there."

"Oh, Gilbert, that would be great. Please do."

"I'll miss you, friend."

"For sure."

We stopped and waited at the pickup stop, and soon a truck appeared from down the road in a cloud of red dust. Geoffrey waved his hand and flagged down the driver.

"I'll see you when school ends," he said. "When you arrive, find someone with a phone and send me their number. We'll talk this way, and I'll make sure Gilbert is there."

"That would be good," I said. "Take care of my windmill, will you? Let me know everything that happens."

"Sure, sure, don't worry."

I squeezed into the pickup with the other passengers, found a sack of charcoal for a seat, and we rolled on to Kasungu. Once there, I caught a minibus down M1 highway to the small town of Madisi. The bus dropped me at a junction on the outskirts of town, where a long road led to the school. I walked a kilometer with my suitcase bouncing behind on the gravel road, until I stood outside the steel gates. In a matter of minutes, I had a dorm room and dorm mates, mealtimes, and a rigorous schedule of classes. Everything was new and foreign and a little overwhelming—but my God, what a pleasure it was to be learning in a real school.

The classrooms in Madisi had solid roofs that didn't leak and smooth concrete floors that didn't have holes. Large unbroken windows let in the sunshine, but kept out the cold. I had an actual desk of my own, complete with pencil holder. During study sessions at night, real fluorescent lights buzzed up above (at least when we weren't having a blackout).

Science class was held in an actual chemistry lab, where the shelves were lined with light microscopes, giant coils of high-resistance wire, glass beakers, and old jars of boric acid. If you can believe it, one of the first lessons our teacher, Mister Precious Kocholola, gave us was about the process of current passing through an electric bell. I'd already applied this concept with my windmill and circuit breaker, but having it explained in scientific terms—in English—was like hearing it for the first time.

But like every other school in Malawi, Madisi relied on the government to survive, and unlike some of the more prestigious boarding schools, it had been neglected. Most of the equipment in the science lab was from President Banda's time, and it was so old that it no longer worked. The chemicals were expired and dangerous, the microscopes rusted and scratched, and for the electric bell lecture, we had no working batteries to supply the power.

"If anyone has an extra dry cell in their rooms, I'll happily demonstrate," the teacher said.

No one did, so we simply used our imaginations.

Our dorms were dirty, and the walls were covered with graffiti. The urinals in the bathroom didn't work, so the Form One students (namely me, the new guy) had to clean them every day to keep down the smell. The rooms themselves were so cramped that two boys were forced to squeeze into one small bed. My bedmate was a boy named Kennedy, who never cleaned his socks.

"*Eh,* man, you need to wash your feet before you come to bed with me," I told him.

"Sorry, I can't ever tell," he said. "I'll wash tomorrow, promise."

But he never did. Often I'd wake up with his feet touching my mouth.

I was several years older than everyone else, so some students started teasing me.

"How many kids did you leave behind at the farm, old man?" they shouted.

"Two boys," I said, "and another one on the way. Perhaps next month."

"The old man thinks he's funny," they said. "You've spent too much time with your cows, herd boy."

One day I decided to end the teasing once and for all. I pulled out the newspaper article about my windmill and put it on the table. "Here," I said. "This is what I was doing."

The boys in my dorm were impressed. "Good job, man!" they said. "How did you manage?" No one teased me after that.

Really, after five years of being a dropout, I was grateful to be in school. But after a couple weeks in this strange place, plus the loneliness from being away from my home and family, I became a little sad. Often after class, I'd hide away in the school library, where the books filled rows and rows of shelves. I'd find a chair and study my lesson books in geography, social studies, agriculture, biology, English, and math. I became lost in American and African history, and within the colorful maps of the world. No matter how foreign and lonely the world was outside, the books always reminded me of home, sitting under the mango tree.

WHILE I ATTENDED SCHOOL at Madisi, Dr. Mchazime was busy making arrangements for the trip to Arusha. Several months before, he'd helped me get a passport. And since I'd never been on an airplane or stayed in a hotel before, he took me out of school one weekend for a crash course on international travel. I took a minibus six hours to Zomba, and we visited the Hotel Masongola, where many tourists stay. He asked the manager to show me a room, how to fill out the guest cards, and how to order in the restaurant. But since the Masongola was too expensive, Dr. Mchazime booked me a room at Peter's Lodge. It was my first night in a hotel, and the first time I'd ever slept on a real mattress.

Dr. Mchazime had also taken a collection to buy me a smart white shirt and black trousers for my journey. They were the nicest clothes I'd ever owned. He also gave me other useful travel advice: for instance, on a plane, I'd be assigned a seat that was mine and mine only, so there was no need to rush and use your elbows like the minibus; if the red light is on near the lavatory, that means it's occupied; and because some people get upset stomachs on their first plane ride, each seat comes with a paper bag for vomit. I was glad to have that bag because I was certain I would need it.

In June, I left school and took a minibus back home to pack. The next morning, a driver appeared to take me to the airport in Lilongwe.

"Our son is leaving us and traveling by airplane," my father said to my mother, smiling.

"That's right," I said. "Flying like a bird in the sky. I'll be waving as I pass over."

"We'll be watching for you. You'll see us here."

My father then tucked a bag of roasted groundnuts in my pocket. They were still warm.

THAT EVENING IN LILONGWE, I was too nervous to sleep, and I stayed up all night in my hotel room watching Super Sport. I was still awake when the sun came up and it was time to leave.

On the plane, I couldn't believe it, but sitting next to me was none other than Soyapi Mumba, the software engineer from Lilongwe who'd first seen my article. Because he's a nice guy, he introduced himself, not knowing who I was. When I said my name and where I was going, he said, "Oh my God, William the windmill guy?" and told me that he'd been the one who'd shown my story to Mike McKay, who blogged about me on Hacktivate. Soyapi was the very reason anybody had even heard about me and my windmill, and the reason why I was going to this conference. Now here he was, sitting next to me on the plane! It also happened that Soyapi was a TED fellow himself, being honored for his coding work with Baobab. I was so happy to find him.

The airplane was bright and clean, and the air-conditioning felt so cool and pleasant on such a hot day. What a great place to be! As the plane taxied toward the runway, I gripped the seat, smiling big. I was certain everyone knew it was my first time. The people seated around me were so well-dressed and confident. They had important things to do, and their busy lives required them to travel in airplanes across the world. As the jet accelerated down the runway and lifted its nose in the air, I pressed my head back in the seat and laughed.

I guess I was now one of them, too.

CHAPTER FIFTEEN

W HEN WE ARRIVED AT the airport in Arusha, Soyapi helped me through customs and immigration, translating for me when my nerves caused my English to disappear. He was staying in a different hotel, so after we got our bags, we parted ways, and I boarded a shuttle bus for the Arusha Hotel. It was dark by the time we left, and I wondered what this new, foreign land would reveal to me in the morning.

The conference was held at the Ngurdoto Mountain Lodge about thirty kilometers outside Arusha. The next morning, pulling out of my hotel, I looked around to see if Tanzania looked and smelled any different than Malawi, but what I saw was very similar: the interstate was filled with minibuses pressed full of people; a giant lorry belched smoke out its back and swerved to miss an old man on a wobbly bicycle. There were children in rags hawking cigarettes on the roadside, while students in bright uniforms marched through the dust to school; I saw village women balancing loads of vegetables on their heads and farmers tending their fields.

But unlike Malawi, Arusha had trees—and not only that. After some minutes, the shuttle driver pointed in the distance and said, "Look there— Kilimanjaro. The biggest mountain in Africa."

There it was, just like in the books, with white clouds covering its top. I couldn't believe that ordinary people like myself climb that great mountain, but I knew they did. When Dr. Mchazime had said there was a great journey ahead of me, I guess he was right. In my head, I began making a list of all the other places in the world I wanted to see.

Kilimanjaro had filled me with great confidence, but it all seemed to vanish once I reached the hotel where the conference was being held. The lobby was filled with so many different kinds of people—lots of white people from Europe and America. There were also many Africans among them, but even they spoke quickly and with strange accents. Everyone was talking on their mobile phones, and I prayed that no one would speak to me. After registering at the welcome center, I walked to the corner of the room and tried to disappear.

No such luck. After some minutes, a man walked up and stuck out his hand. He had red hair and wore bright purple and green glasses.

"Hello, welcome to TED," he said. "My name is Tom. Who are you?"

I'd practiced only one line of English, so I let it fly: "I'm William Kamkwamba, and I'm from Malawi."

He stared at me strangely. Maybe I'd said it in Chichewa.

"Wait a minute," he said. "You're the guy with the windmill."

Tom Rielly was in charge of organizing all the corporate sponsors at TED, including the ones who'd paid for my airfare and hotel. Months earlier at the TED offices in New York, Emeka—the Nigerian blogger—had told Tom about my windmill, saying, "You'll never believe this story . . ." But Tom didn't know that Emeka had then searched under every rock in Malawi to find me and bring me to Arusha. After some minutes—me struggling with the few words of English I knew—Tom asked if I wanted to tell my story on stage to all the "TEDsters," as he called them. I shrugged. Why not?

"Do you have a computer?" he asked.

I shook my head. "No, I don't have."

"Do you have any photos of the windmill?"

I did have these. A friend of Dr. Mchazime's had visited Madisi a few weeks before and helped prepare a presentation, just in case, using photos supplied by the journalists who'd visited my house. He'd done this with a laptop he'd pulled out of his bag—though at the time I had no idea this was a computer. To me, computers were big like televisions and plugged into the wall. We had a few of these at Madisi from President Banda's time, but none of them worked. Before he left, this man handed me a strange cube (a flash drive) attached to a rope and said, "Wear this around your neck. This is your presentation." So when Tom asked about my photos, I unbuttoned my shirt to pull out the rope. Tom looked at me funny, then took the cube and plugged it into his own laptop.

"I'll just copy these onto my computer," he said.

It was then I realized what a laptop was. *Of course,* I thought. *It's a portable computer. What a good idea!*

Sensing my delight at seeing his laptop, Tom asked me, "William, have you ever seen the Internet?"

"No."

In a quiet conference room, Tom sat me down at his computer and explained the track pad, how the motion of my fingers guided the arrow on the screen.

"This is Google," he said. "You can find answers to anything. What do you want to search for?"

"Windmill."

In one second, he'd pulled up five million page results—pictures and models of windmills I'd never even imagined. We did the same for solar power. Next, we pulled up a map of Malawi on Google Earth, then a photo of Wimbe itself, taken from a camera in outer space. It's funny to me now—at this conference in East Africa, with some of the world's greatest minds in science and technology just outside the door, there I was in this room seeing the Internet for the first time. They could have put a blinking sign over my head and charged admission.

Tom then helped me set up my own e-mail account, even sending me a message from another computer to demonstrate. Over the next two days, I'd be introduced to so many amazing pieces of technology, things like BlackBerrys, video and digital cameras, even an iPod Nano, which

I turned over and over in my hand before finally asking, "Where is its battery?" (Not long after, I'd be hacking into these iPods and repairing them.)

But the most amazing thing about TED wasn't the Internet, the gadgets, or even the breakfast buffets with three kinds of meat, plus eggs and pastries and fruits that I dreamed about each night. It was the other Africans who stood onstage each day and shared their stories and vision of how to make our continent a better place for our people.

There was Corneille Ewango, a botanist from Congo who'd risked his life to save endangered animals during the war; he'd even buried his Land Rover engines and stashed lab equipment in the trees to hide them from rebels. A man from Ethiopia invented a kind of refrigerator that works using water evaporation from sand to use in villages without power. A Nigerian named Bola Olabisi started a group to bring together all the women inventors of Africa. Other "TEDsters" were doctors and scientists using creative ideas and methods to fight AIDS, malaria, and tuberculosis. Even Erik Hersman was there—one of the first people, along with Mike McKay, to write about my windmill on his blog Afrigadget. Erik wasn't a biological African (he *was* raised in Kenya and Sudan), but what he said summed up our crowd perfectly:

"Africans bend what little they have to their will every day. Using creativity, they overcome Africa's challenges. Where the world sees trash, Africa recycles. Where the world sees junk, Africa sees rebirth."

Tom helped me prepare the words to say during my own presentation, but of course, they flew out of my head the minute I walked inside the hall. The PowerPoint deck I'd brought with me was a little too long, so Chris Anderson, the TED curator, was going to ask me a few questions onstage instead. By the time I heard Chris call my name, my legs refused to work.

"Don't worry," Tom whispered, patting my shoulder. "Just take a deep breath."

My heart beat fast like a *mganda* drum as I climbed the steps to face the audience, which totaled about four hundred fifty people—among them all the inventors and scientists and doctors who'd stood on that stage in the previous days. They were all in their seats now, watching me. When I walked up the steps and turned around, I went totally blind. Bright lights from the ceiling were shining into my eyes, so bright I couldn't think. All the words I'd prepared seemed to dance to the drum and get lost in the glare.

"We've got a picture," said Chris. He pointed to something behind me, and instantly a giant photo of my parents' house appeared on the wall. I saw the mud-brick walls, grass roof, bright blue sky. I could practically feel the sun.

"Where is this?" he asked.

"This is my home. This is where I live."

"Where? What country?"

"In Malawi, Kasungu," I said. *No, that's wrong.* I quickly corrected myself. "Ah, Kasungu, Malawi," I added. My hands started to shake.

"Five years ago you had an idea," Chris said. "What was that?"

"I want to made a windmill." *Wrong again.* Chris smiled.

"So what did you do, how did you realize that?"

I took a deep breath and gave it my best. "After I drop out from school, I went to library . . . and I get information about windmill . . ." *Keep going, keep going . . .* "And I try, and I made it."

I expected the audience to laugh at my silly English, but to my surprise, all I heard was applause. Not only were people clapping, but they stood up in their seats and cheered. And when I finally returned to my chair, I noticed that several of them were even crying. After all those years of trouble—the famine and constant fear for my family, dropping out of school and my father's grief, Khamba's death, and the teasing I received trying to develop an idea—after all that, I was finally being recognized. For the first time in my life, I felt I was surrounded by people who understood what I did. A great weight seemed to leave my chest and fall to the assembly hall floor. I could finally relax. I was now among colleagues.

For the next couple of days, they lined up to meet me.

"William, can I take my photo with you?"

"William, please join us for lunch!"

One line from my presentation even became a kind of motto for the conference. Everywhere I went, people were shouting, "I try, and I made it!" I was so flattered. I wished my parents, Gilbert, and Geoffrey had been there to see it; they'd have been so proud.

SOMETHING ABOUT MY STORY seemed to tug at Tom's heart. He later told me that as a boy, he'd also spent much of his time tinkering with electronics and dreaming up experiments. When I first met him, he asked what I hoped to obtain someday in my life. I told him I had two goals: to remain in school and to build a bigger windmill to irrigate my family's crops, so we'd never go hungry again. Such a request seemed impossible by Malawian standards, and most people in my country spend entire lifetimes watching such dreams fade. But with the power and influence of the TED community all gathered in Arusha, Tom concluded that school fees and a windmill were relatively simple things to ask for. Tom suggested that since I was a budding entrepreneur—with a good Power-Point presentation, no less—we should try and raise money to achieve those goals.

"You're like a Silicon Valley startup, and I'm going to be on your board of directors," he said. "Let's take this presentation and show it around. We're going to get you some money."

I didn't know anything about Silicon Valley, but I was willing to let him help. For the rest of the conference, with my presentation on his laptop, Tom approached many American investors and business leaders and asked them to help with my projects. He cornered them at dinner and followed them back to their hotels in the shuttle, standing up along the potholed African roads to tell my story. Almost everyone he approached agreed to help; some even opened their wallets right there and handed us hundred-dollar bills.

John Doerr, one of the most successful venture capitalists in the world, volunteered to be one of my first investors. Others such as John Gage, chief scientist from Sun Microsystems, and Jay Walker, a fellow inventor and founder of Priceline.com, later agreed to chip in. I'm so grateful to these kind people and pray that God blesses them all.

After the conference, instead of returning home, Tom flew back to Malawi to help get me enrolled in a better school and to purchase some of the materials I needed to expand the windmill. One of the first things we did in Lilongwe was to buy two mobile phones, one for me and one for my parents, so we could communicate while I was away from home. This way I'd never feel lonely.

Along with Dr. Mchazime, we took a taxi to the village and met my family. When we turned onto the dirt road toward my home, the windmill appeared in the distance, looking so beautiful. As usual, its blades were spinning fast and causing the tower to sway back and forth. Tom stood at the bottom for a long time, taking photos and staring up at it.

"It's more than functional," he said. "William, this is art."

I gave him a tour of the compound, showing him the car battery and bulbs. He laughed at the pile of radio and tractor parts in the corner of my room. "I think every great inventor has a junk pile someplace," he said. I also demonstrated the light switches, circuit breaker, and the way I'd waterproofed my bulb outside. For the porch light, all I'd had was a small car light, so I'd hollowed out a regular incandescent bulb and wired the car light inside. This shell served as both a weather protector and diffuser.

"There's more to this than I thought," Tom said.

I just laughed. I hadn't even told him about the famine.

Back in Lilongwe, Tom and I visited the offices of Baobab Health, located on the grounds of Kamuzu Central Hospital, to see Soyapi and finally meet Mike McKay. Baobab was founded in 2000 by a British-Canadian computer scientist named Gerry Douglas. He'd once been a volunteer with the Ministry of Health and noticed the inefficient methods still used to collect information. Patients were registered by pencil in a big dusty book, which made it nearly impossible to retrieve medical records and compile statistics. Sick people often waited four hours in line just to register and see the doctor. With no easy solutions to this problem, Gerry made his own.

Back home in Pittsburgh, where he lived most of the year with his family, he was on eBay and came across a warehouse full of discontinued iOpener computers—small, cheap units with their hardware built into the screen panels. Gerry initially bought two hundred for twenty dollars apiece, then hacked them into touch-screen systems. He stuck them inside desks with wheels and powered them with car batteries. Over time, he created software that allowed even the most poorly trained hospital staff to scan a bar code and register patients in under a minute. The screen also showed patients' medical histories and how to prescribe their medication, a system that did wonders in giving antiretroviral treatment to people with AIDS. The system's technology, and the efficiency it provided, was in many ways more advanced than anything being used in hospitals in America.

Gerry was in Pittsburgh at the time of our visit, so Mike, Soyapi, and Peter Chirombo, Baobab's hardware technician, gave us the tour. Mike happened to have a small windmill he'd built using an article in *Make Magazine* (now my favorite publication), which he was considering using to power a rural clinic. Its generator was a treadmill motor, which I'd never seen. Peter stuck a power drill in one end of the motor to make it spin, then took the two wires and attached them to a voltmeter—an amazing gadget! Using the voltmeter, I measured the motor's power at forty-eight volts, which was four times stronger than my dynamo.

"What do you think?" Mike asked.

"Yah, it's cool," I said.

He then gave them both to me as a gift. The hole in heaven just seemed to be getting bigger.

Mike and Soyapi also taught me about deep-cycle batteries, which, compared to my car battery, provide a more stable amount of current for longer periods of time. I wanted to try one of these. So Tom and I went to the offices of Solair, a local solar-power dealer, and bought two batteries, four solar LED lamps, along with energy-saving fluorescent bulbs and materials to rewire my entire compound.

Workers came to my village the next week, and for three whole days, we replaced the old wire, dug trenches to bury the cables, and installed proper light fixtures and plugs (though I kept my old flip-flop switches just for show). With new wire, plastic conduit, and buried lines, we never had to worry about starting fires again. I also stuck a lightning rod on top of the windmill just in case. Once finished, there was a bulb for every room, including two outside. Since the dark mud walls of our homes absorbed so much of the light, we painted them white for better reflection. I also installed solar panels on my roof to help supplement the power input. Eventually, every home in my village would have one of these panels, complete with a battery to store power. With each home lighted, the compound glowed at night.

AFTER APPROACHING SEVERAL PRIVATE secondary schools in the area and being turned down because of my old age, I was finally accepted at African Bible College Christian Academy (ABCCA) in Lilongwe, which was run by Presbyterian missionaries. The headmaster, Chuck Wilson, was an American from California, and my teacher, Lorilee MacLean, was from Canada.

Although I was behind most other high school students, Mrs. MacLean and Mister Wilson agreed to take a chance on admitting me. Mrs. MacLean had one condition: that when I left school each day, I didn't go home to poverty. I had to find a good place to live in Lilongwe.

Since I didn't have any relatives in town at the time, Gerry offered me a room at his place. At Gerry's, I had my own bedroom and a desk for studying. The housekeeper, Nancy, also made sure I ate plenty of *nsima* and relish so I wouldn't feel homesick. Everything was great, but because we were in the city, we experienced power cuts several times a week. I couldn't help but think that after all that hardship to bring electricity to my village, here I was sitting in the dark in my success. Gerry said I should bring a windmill with me wherever I go.

Over time, Gerry became a great friend and teacher. He'd been a recreational pilot in England, then later worked on helicopters while living in Canada, so I always had many questions about engines and such. Sometimes after dinner, Gerry explained how helicopters worked, how the spinning blades captured wind to lift the heavy machines, and how the back rotors kept them from spinning in circles. He also helped me with my English, particularly my *l*'s and *r*'s—something we Chichewa speakers always get confused. These lessons were sometimes done in front of the bathroom mirror, so Gerry could demonstrate.

"Okay, William, watch my tongue and say: 'library.'"

"Liblaly."

"L-i-b-R-a-R-y."

"L-i-b-L-a-L-y."

"You'll get it."

My class at ABCCA used a distance-learning curriculum from America that we learned by computer over the Internet. Just a couple months before I'd never even seen the Web, and now I used it every day to speak with teachers who were in Colorado. My high school class was small, only twelve people, including two Americans, a Canadian, a Korean, and a boy and girl from Ethiopia. A lot of Malawian children are in the primary school at ABCCA, but I was the only local high school student (the tuition was five thousand dollars per year, which most Malawians can't afford). At first, I was a bit ashamed of my poor English, especially after hearing five-year-old children speak better sentences than I could. During my first few days, I became a little depressed. But my tutor, a Malawian named Blessings Chikakula, offered some great encouragement.

Mister Blessings had also come from a poor village near Dowa, where he'd worked as a primary school teacher, supporting his wife and four children on almost no pay. During the famine, the village suffered terribly and people died, including his father and several of his students. So, desperate to feed his family and give them something better, he'd caught a minibus to Lilongwe to join the Malawian army. But just as he was about to enter the gates, he received a call from his cousin, who told him that a scholarship application Blessings had submitted months before to ABCCA had been accepted. Later, when he was thirty years old, Blessings had stood proud before his wife and four children and walked across the graduation stage. The school then hired him as a teacher.

Me and my parents standing next to my windmill in 2007, just after I attended the TED conference.

"Don't be discouraged and give up just because it's hard," Blessings told me. "Look at me. I didn't go to college until I was thirty. Whatever you want to do, if you do it with all your heart, it will happen."

Eventually, the money from my donors allowed me to help my family in many other ways. I installed iron sheets on every relative's home in our village to replace the grass thatch. I got mattresses so my sisters no longer had to sleep on grass mats on the dirt floor, plus covered water buckets to keep our drinking supply safe from pests. I bought better blankets to keep us warm at night in winter, malaria pills and mosquito nets for the rainy season, and I arranged to send everyone in my family to the doctor and dentist. (After never having seen a dentist in my life, I had only one cavity!)

And for once, I finally managed to repay Gilbert for all the help he'd given me. After Gilbert's father died, Gilbert had to drop out of school because the family couldn't afford to send him. So with my donations, I put Gilbert back in school, along with Geoffrey and several other cousins who'd dropped out during the famine. I even sent the neighbors' kids back to school.

And after years of dreaming about it, I was finally able to drill a borehole for a deep well, which gave my family clean drinking water. My mother said this saved her two hours each day carrying water from the public well. Using a solar-powered pump, I filled two five-thousand-liter water tanks and piped them to my father's field. I then installed easy-assembly drip lines purchased from an American company called Chapin Living Waters, which finally allowed my father to plant a second maize crop. The storage room will never be empty again. The spigot from the borehole—the only automatic water system for miles around—is also free for all the women in Wimbe to use. Each day, dozens of them come to my home to fill their buckets with clean, cool water without having to pump and pump.

During holidays off school from ABCCA, I constructed a bigger windmill that pumped water. That pump now sits above the shallow well at home and irrigates a garden where my mother grows spinach, carrots, tomatoes, and Irish potatoes, both for my family to eat and to sell at market. Finally, the dream had been realized.

My family couldn't have imagined that the little windmill I built during the famine would change their lives in every way, and they saw this change as a gift from heaven. Whenever I came home from school on weekends, my parents had a new nickname. They called me Noah—like the man in the Bible who built the ark, saving his family from God's flood.

"Everyone laughed at Noah, but look what happened," my mother said. "He saved his family from destruction."

"You've put us on the map," my father said. "Now the world knows we're here."

In December 2007, I went to the United States to visit Tom for the Christmas holidays and to see the windmills of Southern California that appeared in my book. After some difficulty getting a visa—it's never easy for us Africans to travel to Europe or America—I landed in New York City right in the middle of winter, wearing only a sweater. The woman at the airline counter then told me that they'd lost all my luggage.

"We'll call you," she said.

I wondered how. I didn't even have a phone.

In the taxi from the airport, I finally saw the great American infrastructure I'd read so much about. We drove over smooth roads with five lanes in each direction, across bridges with no water underneath, followed by more roads and more bridges. The tall buildings in the distance appeared so strong and pressed together. It was hard to imagine that men had built these things and a person could even walk between them.

It just so happened that Tom lived in one of those buildings in lower Manhattan. His apartment was on the thirty-sixth floor, and I wondered how I would ever get up there. Someone then showed me the elevator, which took me up in ten seconds just by pushing a button. Looking up, I saw that there was even a mirror on the ceiling. Already, I had so many questions.

Inside, the apartment was surrounded in windows from floor to ceiling, looking as if you could walk right over the edge. Before that day, the highest I'd ever been was the top of my windmill. It took me some time to adjust, and that night on the sofa, I had trouble sleeping.

Tom had to work a couple of days that week so several of his friends volunteered to show me around the city. One of them had arranged for a pile of warm clothes to be waiting when I arrived from the airport— complete with a winter coat, gloves, a scarf, and a hat so I wouldn't freeze. I was so grateful, especially because everything I owned was lost someplace in Africa.

The next day, another friend—a famous dancer named Monica Gillette—picked me up for a sightseeing tour of Manhattan. Down in the subway, I watched people enter the gates using sliding cards—what a great idea. Traveling through the tunnels with the giant buildings above our heads, I was amazed that they never fell on us. The sidewalks of New York left me exhausted, with hundreds of people running in every direction. One of the things I noticed in New York is that people don't have time for anything, not even to sit down for coffee—instead, they drink it from paper cups while they walk and send e-mails.

Standing at a construction site, I watched giant cranes lift enormous pieces of steel into the sky, and it made me wonder how Americans could build these skyscrapers in a year, but in four decades of independence, Malawi can't even pipe clean water to a village. We can send witch planes into the skies and ghost trucks along the roads, but we can't even keep electricity in our homes. We always seem to be struggling to catch up. Even with so many smart and hardworking people, we were still living and dying like our ancestors.

The nice lady who'd donated my warm winter clothes was named Andrea Barthello, and she and her husband, Bill Ritchie, run a company called ThinkFun that makes cool educational toys and puzzles. I'd met her at TED, along with their son Sam, who was my age. The next day, Andrea and Sam took me on another tour of Manhattan island—this time by helicopter! Inside the chopper, the pilot let me wear headphones and sit up front near the instrument panels, dials, switches, and screens. Our glass bubble took us high above the city, above the Statue of Liberty, past the Empire State Building and speedboats on the river leaving their long trails of white. I couldn't stop grinning.

Later that week, Tom and I drove to Connecticut to have dinner with Jay Walker and his wife, Eileen, whom I'd met at the TED conference. The library in Jay's home is famous throughout the world, a kind of museum of great inventions. Many of his books were hundreds of years old and written on whatever materials were available, showing how even in other parts of the world, people went to great trouble to expand their knowledge. Some of the books were even covered in jewels. But the library was more than just books: an original Soviet Sputnik satellite hung from the ceiling, and displayed on a shelf were some of the first computers, radios, and processors, even a Nazi Enigma code machine. But my favorite item was a replica of Thomas Edison's first lightbulb, which I studied from every angle. I'd had a hard enough time creating light from a windmill. But *my God,* this man had created the actual light!

To think, my journey had begun in my tiny library at Wimbe—its three shelves of books like my entire universe. But now standing here, I was seeing the true size of the world, and how little access I had to it. There was so much to see and do. I felt a bit light-headed.

After several days, we flew to California and spent Christmas Day in Los Angeles with Tom's sister and family. We drove to the Santa Monica pier on the Pacific Ocean and watched the surfers. Along the boardwalk in Venice Beach, I spoke to a man who blew fire from his mouth and walked on broken glass.

At the San Diego Wild Animal Park, which we visited later that week, I saw giraffes, hippos, elephants, rhinos, monkeys, and more. You know, half an hour east of my village was Kasungu National Park, and a couple hours west was Nkhotakota Wildlife Reserve, where many of those same animals roamed and played. I'd never been to either place. Instead, I'd traveled ten thousand miles to see them in America. This made me laugh.

Early the next day, we drove through the desert to Las Vegas for the Consumer Electronics Show, which was an endless maze of toys and cool gadgets, many of them using low-power technology, such as solar-powered LED lamps, a Bluetooth headset that hooks over your ear and recharges your phone from the sun, and Freeplay radios powered by a crank that never need batteries. That night, we stayed in Treasure Island Hotel, where jackpots of cars and other prizes are given away all night long and women in their underpants serve free soda.

Standing there on the casino floor amid the blinking lights, as fountains of money spilled from the machines to the sound of sirens, I had to remind myself to breathe.

"You having fun?" Tom asked, shouting over the noise.

"Yah," I said, smiling big. "Great."

But really, all the stimulation from the past week was too much for my brain to process, so I drifted back to the one place I knew. I stood there at the base of my machine, then scaled the tower rungs—slowly, one by one, feeling the soft wood creak beneath my bare feet, so smooth and warm in the sun. From the top, I looked out onto the country I loved—across the vast green fields and craggy slopes of the highlands, which sent a familiar breeze through the valley and whipped the blades behind me. Later that night as I lay in bed, I let that daydream spin me off to sleep, the white noise of the machinery like a song my mother would sing. I went to sleep dreaming of Malawi, and all the things made possible when your dreams are powered by your heart.

THAT VISION OF HOME was a place in my mind where I retreated often while traveling in America. It was always the same, and each time I went there, it brought a warm, pleasant feeling.

But not long after that night in Vegas, I found myself staring out onto a near-identical landscape of home as my dream—green rolling flatland, mountains on the horizon, draped in bright blue sky. Except I wasn't in Malawi—I was in Palm Springs, California. And in that empty space between me and the hills, there suddenly appeared miles of windmills, more than six thousand of them, shooting from the earth like a giant forest of mechanical trees.

The true dimensions of the machines revealed themselves once we pulled into one of the wind farms. There, I seemed to lose my sense of scale. The round white trunks of the turbines were like cartoons I'd seen on television, so wide they could swallow my family's entire house. Stepping outside, I was greeted by the sound of rushing wind, so deep and encompassing, it seemed to pull my very breath. Looking up, I saw the hundred-foot blades twirling slowly like the toys of God.

I had to be dreaming this.

These windmills were different from mine in every way. Each machine towered more than two hundred feet in the air, with wingspans longer than the airplane that had brought me to America. The head engineer of the Windtec wind farm, Chris Copeland, even took me inside the dark belly of one of the machines. On one wall, a computer monitor told us all kinds of information—everything from voltage output to wind and blade speed. If the gusts became too strong, the computers shut down the rotors—unlike my windmill at home that sometimes snapped apart in heavy wind, sending the blades spinning through the air like flying knives.

Each turbine generated 12,500 volts, and in total, the wind farm produced over six hundred megawatts, which was delivered to a substation by underground cable, then to thousands of homes in Southern California. Six hundred megawatts was enough to power the entire country of Malawi, with energy to spare (ESCOM, by comparison, produced only 224 megawatts).

It was an incredible feeling to see the machines that I'd been imagining for so long. Now here they were, twisting in the wind before me. I'd come full circle. The pictures in the library book had provided the idea, hunger and darkness had given me the inspiration, and I'd set out myself on this long, amazing journey. Standing there, I waited for the next direction. *What will I do next?* I thought. What was in my future, after having come this far? I looked out across the expanse of machines and saw how the mountains seemed to tumble and dance along their twirling blades.

As I watched them, they seemed to be telling me something—that I didn't have to decide just then. I could return to Africa and go back to school, reclaim the life that had been taken from me for so long. And after that, who knows? Perhaps I would study these machines and learn how to build them, then plant my own forest of them along the green fields of Malawi. Perhaps I would teach others to build more simple windmills like the one I had at home, to provide their own light and water without having to depend on the government. Perhaps I'd do both. But whatever it was I decided to do, I would apply this one lesson I'd learned:

If you want to make it, all you have to do is try.

EPILOGUE

*I*N JUNE 2008, I traveled to Cape Town, South Africa, for the World Economic Forum on Africa, and spoke about technology in emerging countries. I was part of a panel discussion moderated by Dan Shine from Advanced Micro Devices' 50x15 Initiative, which is working to connect 50 percent of the world's population to the Internet by 2015. I'd met Dan in Arusha and he'd become my friend, so when he asked me to speak at this important event, I happily said yes.

I told the audience how I'd built my windmill and also about the different challenges I had faced, particularly when it came to finding a generator. I told them about Geoffrey, and that I hoped he could begin teaching people in the surrounding villages to build windmills while I was away at school. At the end, someone in the audience raised their hand and asked me what the Malawian government thought about my project.

"The government doesn't know about it," I said. I knew this because I'd just informed the president myself.

Malawian president Bingu wa Mutharika was also attending the conference, and the night before, we were at the same dinner reception. I admired President Mutharika because he cared about us farmers and made it possible for families like mine to afford fertilizer once again. The president was seated at another table, and after the dinner, I walked right up and introduced myself—in English. I told him I was a speaker at the forum and that I'd been invited because of my windmill. The president looked surprised.

"Oh, that's good news!" he said, then stood by me and posed for a photo. I was so proud. That picture now hangs on the wall in my parents' living room and they make sure every visitor admires it.

AFTER LEAVING CAPE TOWN, I flew to Chicago, where I was being honored at the Museum of Science and Industry. I was part of an exhibit called "Fast Forward: Inventing the Future" that highlighted some of the innovative technology and ideas that will shape our world for the better. My original circuit breaker and light switch were there, sitting alongside exhibits celebrating the work of people like Ayanna Howard, a robotics engineer who worked on NASA's SmartNav Mars rover, and Dana Myers, who invented an all-electric car that can travel at 70 miles per hour. It was such an amazing honor to be mentioned alongside such smart people, and seeing my face blown up on a poster almost as big as my body was also pretty cool, if not kind of strange.

Once I got back home to Malawi, I spent the summer seeing my family and friends and did some much-needed repairs on my windmills. Every time I go home, it seems that one of my blades has snapped off from strong winds. This continued to happen even after I replaced the plastic blades with steel ones from an oil drum. I also noticed termites had burrowed through the wooden legs of my tower, right at the base, meaning I'd have to rebuild the entire thing (I later planted the legs in cement). With wobbly, termite-eaten wood, the tower legs became even more dangerous to climb when it was time to make repairs. Sometimes I wore a bicycle helmet in case I fell on my head.

WHILE I WAS TRAVELING in America over Christmas, I'd received news that I'd been accepted—with a scholarship—to the African

Leadership Academy, a pan-African high school in Johannesburg, South Africa.

The school brings together students from fifty-three countries with a mission to train Africa's next generation of leaders. Out of 1,700 applicants, only 106 were accepted for its inaugural year. Many of these students are inventors and entrepreneurs like me who've overcome difficult odds to better the lives of their families and neighbors. Others are simply some of the smartest students in their countries, scoring the highest marks on their national exams.

Despite having worked very hard in Lilongwe at my previous school, I was still behind in English and math. I knew the school in Johannesburg would be extremely challenging, and I feared other students would be far more advanced (and younger) than me. Since speaking English was one of my biggest concerns, one of my American sponsors volunteered to send me to a language class in Cambridge, England. For five weeks, I attended classes near the Cambridge University campus and studied the Queen's English along with students from China, Italy, and Turkey.

On the weekends, I walked the old city and learned about its buildings, many of them built by hand more than four hundred years ago without the kind of modern technology we have today. Seeing this, it gave me even more confidence that we Africans can develop our continent if we just put our minds and abundant resources together and stop waiting on others to do it for us.

IN AUGUST, AFTER COMING home briefly to pack and say good-bye to my family once again, I boarded a plane and flew to Johannesburg. My classes at ALA were just as rigorous as I'd expected. I took ten courses my first semester, including chemistry, physics, and a class in entrepreneurship, which became my favorite. The school is located outside Johannesburg on a beautiful campus with giant trees; wide, green soccer pitches; and lots of peacocks, which are even louder and more disturbing in the morning than the noisy chickens at home. I share a room with a Kenyan guy named

Githiora Thuku, who quickly became a good pal. But Githiora isn't forced to share my bed, and anyway, I'm pretty sure he washes his feet more frequently than my previous roommate.

For the first time since the TED conference, I felt like I was surrounded by real colleagues who shared the same motivations—but this time in a much deeper way, because we'd also shared similar hardships in getting to this place.

There's Miranda Nyathi from KwaZakhele, South Africa. During a big teachers' strike that paralyzed her school, Miranda took charge and began teaching her fellow students math, science, and geography and managed to keep the school open. And Belinda Munemo from Harare, Zimbabwe, helped a local orphan girl start a successful hatchery and chicken farm to pay for her school fees. Since then, Miranda has opened a small video store so students can rent educational videos for free instead of wasting away in the boozing centers. My friend Paul Lorem is a "Lost Boy" from southern Sudan who survived the war and lived alone without parents in a refugee camp, much like my other friend Joseph Munyambanza from Congo, whose family fled the fighting and lived in a camp in Uganda, where Joseph attended school.

All these people's stories are so inspiring to me. Even when my classes become difficult and I find myself discouraged, just being around them helps me continue. And I wonder just how many others like us are still out there struggling on their journey. Thinking of them reminds me of a quote I read recently from the great Reverend Martin Luther King Jr. that says, "If you can't fly, run; if you can't run, walk; if you can't walk, crawl." We must encourage those still struggling to keep moving forward. My fellow students and I talk about creating a new kind of Africa, a place of leaders instead of victims, a home of innovation rather than charity. I hope this story finds its way to our brothers and sisters out there who are trying to elevate themselves and their communities, but who may feel discouraged by their poor situation. I want them to know they're not alone. By working together, we can help remove this burden of bad luck from their backs, just as I did, and use it to build a better future.

ACKNOWLEDGMENTS

William Kamkwamba: First I'd like to thank my parents and sisters and all those in my family, especially my cousin Geoffrey. Thanks to Gilbert Mofat, who was always by my side and became my partner in everything. I wouldn't be where I am today had it not been for the diligence of Dr. Hartford Mchazime at the Malawi Teacher Training Activity, who will always hold a special place in my heart. Special gratitude to Blessings Chikakula, a wonderful tutor and teacher, and now a dear and loyal friend. Thanks to Eneret Santhe and Ralph Kathewala at MTTA; to Frederick M. K. T. Mwale, Carlos Lucas Chimuti, Edith Sikelo, and Gloria Mlokowa at Wimbe Primary School; Rhonex Banda at Madisi Secondary School; and McDonald Tembo at Kachokolo Secondary School. And thanks to the rest of my brothers and sisters in the great Republic of Malawi, for we've struggled together and we'll succeed together, and your hard work and toughness make me proud of who I am. (Uncle John, Grandpa Matiki, Chief Wimbe, and Khamba—may you all rest in peace!)

Attending the TED conference changed my life and sent me on the new exciting journey that I'm on today. Much of that wouldn't be possible without the assistance of Tom Rielly, who's helped open many doors of opportunity, in addition to clearing the goats from the road and pointing me in the right direction. Along the way, Tom has become a trusted mentor and friend, and to him I'm always grateful. The rest of the TED community continues to provide endless inspiration. Special thanks to Chris Anderson, June Cohen, and Emeka Okafor, and to the wonderful people who've invested in my future out of the kindness of their hearts: John Doerr, Mike and Jackie Bezos, Jay and Eileen Walker, John Gage, Gerry Ohrstrom, Andrea Barthello, and Bill and Sam Ritchie. You've all elevated my life in more ways than you know.

Thanks to Soyapi Mumba and Mike McKay for first sharing my story with the world, and to Gerry Douglas for giving me a home in Lilongwe and being a friend. (Also to Nancy Nyani for being my second mother in the city.) Thanks to Chuck and Joanne Wilson and Lorilee MacLean at African Bible College Christian Academy, and to Fred Swaniker, Chris Bradford, Christopher Khaemba, Scott Rubin, Gavin Peter, David Scudder, and Alison Rodseth at African Leadership Academy—I'm grateful to all of you for believing in my potential as a student and giving me the invaluable opportunity of a good education.

Thanks to my cool New York City documentary crew: Ben Nabors, Michael Tyburski, Scott Thrift, and Ari Kuschnir; to Sarah Childress at the *Wall Street Journal*; and to Monica Gillette, Will Allen, Matt Curtis, and Perrin Drumm. Thanks to Dan Shine at Advanced Micro Devices and Kathleen McCarthy and Cynthia Morgan at Chicago's Museum of Science and Industry. Thanks to my agent, Heather Schroder, at International Creative Management; to my editor, Henry Ferris, and deputy publisher Lynn Grady at William Morrow; and to Seale Ballinger and all the kind people at HarperCollins who've made me feel so welcome.

And last, thanks to my coauthor, Bryan Mealer, who gave me a voice. Without his incredible hard work this book would have literally been impossible. I've learned so much from him, and I'm proud to call him a part of my family. He is my brother.

Bryan Mealer: First of all, thanks to William for never giving up and for allowing me to help him share his uplifting story with the world, and to my agent, Heather Schroder, for making the introduction. This book wouldn't have been possible without the unconditional cooperation and assistance of William and his parents, Trywell and Agnes Kamkwamba, who took me in for months, fed me, and answered every question I asked. In the process, I became part of their amazing family and gained three incredible friends. Blessings Chikakula spent nearly every day with me in Malawi, acting as both cultural interpreter and one of the best translators I've ever worked with. He's a gifted educator and a proud African. I'll know him for the rest of my life. Thanks to Gilbert and Geoffrey for letting me hang out, to Gerry Douglas for giving me a bed in Lilongwe, and to Nancy Nyani for her wonderful meals and company. And special thanks to Tom Rielly for his bottomless generosity, and for consistently going above and beyond to help me tell this story.

I'm also grateful to the kind people at African Leadership Academy, especially Christopher Khaemba and his lovely family for hosting me. Scott Rubin guided me through marathon sessions on electromagnetic induction and basic physics, all while William looked on and laughed. Brushing up on science was a big part of the reporting, and for that I'd also like to thank Meghan Maresh, Tony Blatnica, and Chris Copeland. Thanks to my editor, Henry Ferris, and all the dedicated folks at William Morrow and HarperCollins who believed in this book. And finally, thanks to my beautiful wife, Ann Marie Healy, who shared in my joy of finally coming home from Africa with some good news to tell.